STEPHEN CANTLE is a qualified Biology teacher and has acted as an examiner for a number of GCE examination boards in O-level Human Biology, Biology, Human and Social Biology as well as the A-level General Studies paper.

After graduating from London University with a degree in Human Biology, he taught Biology in a number of schools. He then studied for a Masters degree in Medical Science at the University of Nottingham, before joining the teaching staff at the Science Museum in London.

He later joined the editorial staff of the new *Times Health Supplement* as a journalist and is now a medical reporter for the weekly newspapers, *Doctor* and *Hospital Doctor,* based in Guildford.

He is author of the Human Biology cards in the Key Facts series, and also of a book of Human Biology multiple-choice questions.

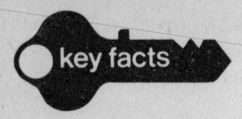

GCE O-Level Passbooks

BIOLOGY, R. J. Whitaker, B.Sc. and
J. M. Kelly, B.Sc., M.I.Biol.

CHEMISTRY, C. W. Lapham, M.Sc.

COMMERCE, L. D. Hunter, M.A.

COMPUTER STUDIES, R. J. Bradley, B.Sc.

ECONOMICS, J. E. Waszek, B.Sc.(Econ.)

ENGLISH LANGUAGE, Robert L. Wilson, M.A.

FRENCH, G. Butler, B.A.

GEOGRAPHY, R. Knowles, M.A.

GEOGRAPHY, BRITISH ISLES,
D. Bryant, B.A. and R. Knowles, M.A.

GERMAN, A. Nockels, M.A.

HISTORY, Political and Constitutional
(1815-1951), L. E. James, B.A., M.Litt.

HISTORY, Social and Economic
(1815-1951), M.C. James, B.A.
and L. E. James, B.A., M.Litt.

HUMAN BIOLOGY, S. Cantle,
B.Sc., M.Med.Sci.

MODERN MATHEMATICS, A. J. Sly, B.A.

PHYSICS, B. P. Brindle, B.Sc.

RELIGIOUS STUDIES, D. Stent, B.Ed.

TECHNICAL DRAWING, P. J. Barnett,
D.S.C., M.C.C.Ed., Adv.Dip.Ed.

GCE O-Level Passbook

Human Biology

Stephen Cantle, B.Sc., M.Med. Sci.

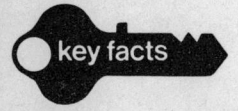

Published by Charles Letts and Co Ltd
London, Edinburgh and New York

This book is sold subject to the condition that it shall not, by way of trade or otherwise, be lent, re-sold, hired out, or otherwise circulated without the publisher's prior consent in any form of binding or cover other than that in which it is published and without a similar condition including this condition being imposed on the subsequent purchaser.

© 1983 Charles Letts & Co Ltd
Diary House, Borough Road, London SE1 1DW
Reprinted 1985
Made and printed by Charles Letts (Scotland) Ltd
ISBN 0 85097 562 X

Contents

Introduction 7

1. Cell Structure and Functioning 9
2. The Gut and Nutrition 39
3. The Lungs and Ventilation 60
4. Blood: the Body's Transport System 72
5. The Kidneys, Skin and Excretion 92
6. Sensitivity and Co-ordination 106
7. The Skeleton, Locomotion and Growth 135
8. Reproduction and Inheritance 150
9. Health and Disease 179
10. Man and Other Living Organisms 204

Examinations and Exam Technique 215

Index 222

Units Used in This Book

S.I. (International System) units are used throughout.

The abbreviations are:

m = metre = 1000 millimetres
mm = millimetre (0.001 m)
μm = micrometre (0.000 001 m)
nm = nanometre (0.000 000 001 m)

g = gram = 1000 milligrams
mg = milligram
J = joule
mm^3 = cubic millimetre

Introduction

This book will help you pass your GCE O Level or CSE examination in Human Biology. It provides an up-to-date, detailed account of all the information required by the British GCE and CSE examination boards that set papers in Human Biology. It is intended as a revision aid to help you prepare for an exam at the end of a course of study, but can also be used as a handy-sized reference book.

After buying the book you should obtain a copy of the syllabus for the particular examination you will be sitting. Ask your teacher to help you get a copy. Then eliminate from the book any section that may not be required by your particular syllabus.

Human Biology is a fascinating subject and is generally very popular among pupils. People usually enjoy finding out how their own body works and why it sometimes goes wrong. For instance, you can find out why you resemble your parents, or how your body responds to the physical stress of exercise. Other topics include the use and abuse of drugs, contraception and human growth. These subjects make Human Biology more immediately interesting than, say, the behaviour of electrons in Physics, or a long list of dates in History.

Human Biology is also important because it forms the scientific basis for health education. People who understand the effects on the body of smoking cigarettes, overeating or the excesses of alcohol, are more likely to avoid their ill-effects. Knowledge of your body also helps when you become unwell. You should know, for example, why you should always take the full course of a treatment with antibiotics, or why you should not drink too many cups of tea or coffee. Human Biology as the basis of first-aid will also help you deal sensibly with other peoples' ill-health or accidents.

Human Biology is a science. This means that facts are obtained by direct observation and experimentation; the so-called scientific method. It is a biological science because its subject of study, man, is a living organism. This means that in common with other living things, man shows all the seven characteristics of animate objects; locomotion (movement); sensitivity (response to stimuli); nutrition (obtaining food); respiration (uses energy); excretion (removal of waste); growth and cell repair; and reproduction.

These basic processes, together with the structure and chemical functioning (and malfunctioning) of the different types of human cells, tissues and organs, are the chief concerns of the human biologist. Along with general ecological aspects of man, such as his dependence on other living organisms, these concerns form the basis of the various CSE, GCE O and AO Level syllabuses in Human Biology.

Key words or concepts are marked in the text by bold type and they are listed, together with their definitions, at the end of each chapter. Comprehensive advice about examination technique can be found at the end of the book. However, no amount of advice will help unless the appropriate examination syllabus has been thoroughly studied, understood and revised. The following pages will help you do the latter.

Chapter 1
Cell Structure and Functioning

The cell is the basic structural and functional unit of the human body. The body of an average-sized person contains about one hundred million million of them (10^{14}). Human cells come in different shapes and sizes according to the function they perform. One of the smallest is the red blood cell, which is less than 10 μm in diameter. One of the largest is the female germ cell, or ovum, which is about 100 μm in diameter. Some cells are spherical, like those from the salivary glands in the mouth. Others such as nerve cells are long and thin, some of them as long as 1 m.

Human or other animal cells differ from plant cells in a number of important respects. These differences can be seen by looking at plant and animal cells under a light microscope (Figure 1), giving a magnification of up to 1,500 times.

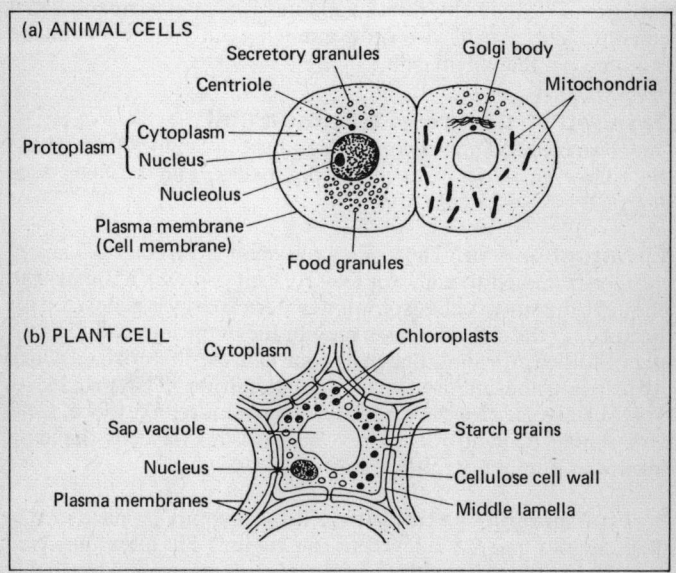

Figure 1. Animal and plant cells (mag. $\times$ 400)

A typical animal cell has a diameter of about 20 μm. It consists of a thin membrane called the **plasma membrane** which encloses the cell's contents, the **protoplasm.** Suspended in the protoplasm is a dark spherical object, the **nucleus,** which is itself bound by a membrane and contains a dense area, the nucleolus. The remainder of the protoplasm is called **cytoplasm.** This contains a number of different granules and structures which can be seen more clearly under an electron microscope.

Most of the structures found in the animal cell are also found in a typical plant cell. But as Figure 1 shows, there are some additional structures as well. The main visible difference is the presence of a large space, or vacuole, in the centre of the cell, which is filled with a solution, the cell sap. The plant cell membrane is thick and made of cellulose. Because of the presence of the cell vacuole, the plant cell's protoplasm is compressed. A nucleus is present, as are a number of new features; granules of starch and spherical bodies known as chloroplasts, which contain the green pigment **chlorophyll.** Chlorophyll is able to absorb light energy from the sun and use it to convert carbon dioxide and water into sugars by the process of **photosynthesis.** Animal cells do not have a cellulose wall, never contain chlorophyll and never have a permanent cell vacuole. Animal cells also show a much greater variety in shape and function than plant cells.

Detailed structure of a human cell

Under an electron microscope (magnification up to 500,00 times) the following structures, called organelles, can be seen (see Figure 2).

1. Mitochondria. These tiny rod-shaped structures vary in number from cell to cell. An average cell contains about 1,000. They are the site of cell respiration, where energy is generated for the needs of the cell. The inner membrane of the mitochondrion is highly folded, forming finger-like structures, the cristae, which project into the middle of the mitochondrion. This provides a large surface area for the respiratory reactions to take place. Cells with a high demand for energy, such as muscle cells, have an enormous number of mitochondria.

2. Endoplasmic reticulum. This is a complex maze of tiny slits, cavities and passageways. The cavities are interconnected and are continuous with the nuclear membrane. The highly folded membranes of the endoplasmic reticulum (ER) are in

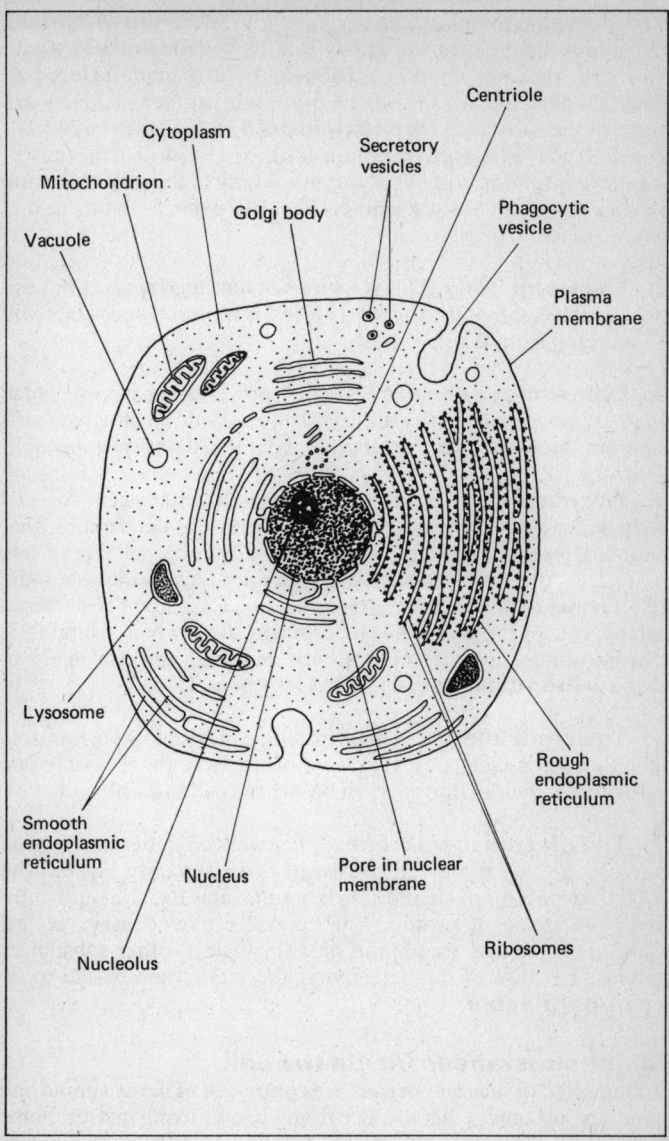

Figure 2. Electron micrograph of an animal cell (mag. × 10,000)

places covered in tiny structures known as **ribosomes.** (Where ribosomes are present the ER is said to be 'rough', and where they are absent, 'smooth'.) Ribosomes are manufactured in the nucleolus and are the site of protein manufacture. They are made of the substance **ribonucleic acid** (RNA) (see page 167). The ER acts as a sort of mini-transport system for the movement of proteins in and out of the cell. Many of the proteins made by a cell are for use elsewhere. The ER exports them, that is secretes them from the cell.

3. The Golgi body. This is a series of flattened spaces, or vesicles, which is thought to act as a store for the substances that will be secreted from the cell.

4. Lysosomes. These are sacs of powerful enzymes, which when released can destroy the cell. They provide an effective 'self-destruct' mechanism for destroying a damaged or unwanted cell.

5. The nucleus. This is the largest structure in the cell. All cells have one at some time in their lives. The nucleus determines what happens inside the cell. It consists of a number of thread-like structures, the **chromosomes,** which are only visible when the cell is about to divide. All normal human cells contain 46 chromosomes, except the germ cells (sperm and ova), which contain 23. Chromosomes are the cell's genetic material and are made of **deoxyribonucleic acid** (DNA) (see page 15).

6. The centriole. This small structure plays an important role in cell division (see page 31). Most of the time there is only one centriole. It divides into two just before the cell as a whole divides.

7. The plasma membrane. When viewed under an electron microscope this is seen to consist of three separate layers. The total thickness of the membrane is about 7 nm. Its inner and outer layer are made of protein, the central region consists of fat. Certain substances are able to pass through it, other substances cannot. Because of this selectivity, the membrane is said to be **semi-permeable.**

Major substances inside the cell

Living cells, tissues and organs are composed of many simple and complex substances that are also found in non-living matter. Some substances are unique to living organisms. The study of substances that occur in living cells and their reactions is the science of biochemistry. The study of their structure and behaviour is known as molecular biology.

Chemical substances are either **organic** or **inorganic**. Organic substances are often thought to be only those that occur in living systems. This is incorrect. They are all the complex compounds that contain the element **carbon**. Organic substances are not unique to living systems; many can be synthesized in the laboratory. Inorganic substances are substances that do not contain carbon (excepting carbon dioxide, carbonic acid, bicarbonate etc.).

The principal organic substances that are found in living organisms are: proteins, carbohydrates, fats and nucleic acids. The principal inorganic ingredients are: water, salts, acids and bases.

1. Proteins. These are large complex substances made up of much simpler substances called **amino acids.** All amino acids contain hydrogen, carbon, oxygen and nitrogen. Some also contain phosphorus and/or sulphur. The simplest amino acid is glycine, which has the following chemical structure:

$$H_2N-CH_2-COOH \quad \text{or} \quad NH_2 . CH_2 . COOH$$

All amino acids contain at least one COOH group and at least one NH_2 group. Amino acids can join together by forming linkages known as **peptide bonds.** This involves loss of a water molecule (condensation) between the COOH group of one amino acid and the NH_2 group of another. Two amino acids joined together by one peptide bond comprise a **dipeptide.** A molecule containing more than one peptide bond is called a **polypeptide.** A polypeptide is therefore a chain of amino acids. Proteins consist of one or more polypeptide chains, sometimes bound to another compound. **Haemoglobin,** the red pigment of the blood, for example, consists of a large, complex protein, **globin,** bound to an iron compound, **haem.**

Proteins such as haemoglobin are manufactured inside the body's cells from amino acids obtained from dietary protein. Amino acids cannot be stored in the body, so a steady supply of them must be supplied by the diet. Surplus amino acids are broken down by the liver, (see page 56), where the NH_2 part of the molecules is converted into **urea** and the remainder used for cell respiration.

Proteins make up about 12 per cent of the weight of the human body and are the main structural components of cells. Muscles in particular contain a high concentration of protein and proteins are also found in large quantities in bone. An important group of proteins is the body's **enzymes,** *all* of which are proteins (see page 167).

2. Carbohydrates. As their name suggests, carbohydrates are hydrates of carbon. They therefore consist of carbon, hydrogen and oxygen only, with twice as much hydrogen as oxygen. The simplest carbohydrate is **glucose,** which is composed of a ring of six carbon atoms with attached hydrogen and oxygen atoms. Its structure is usually represented as follows:

$$\begin{array}{c} H_2C-OH \\ | \\ C-O \\ /| \quad \backslash \\ H \quad H \quad OH \\ |/ \quad | \\ C \quad OH \quad H \quad C \\ |\backslash | \quad |/| \\ OH \quad C-C \quad H \\ | \quad | \\ H \quad OH \end{array} \quad \text{or } C_6H_{12}O_6$$

Simple sugars, such as glucose, fructose and galactose, have only one ring, and are called **monosaccharides.** Like amino acids they can join together to form more complex molecules. Two glucose molecules can join together to form **maltose** by losing one molecule of water. Maltose, $C_{12}H_{22}O_{11}$, and sucrose and lactose, are examples of **disaccharides.** Carbohydrates with more than two glucose rings are called **polysaccharides. Starch,** for example, is a long chain of several hundred glucose molecules. Another polysaccharide is **cellulose,** the main constituent of plant cell walls.

Simple sugars such as glucose, fructose, sucrose and maltose are all soluble in water and are sweet-tasting. Polysaccharides such as starch, cellulose and **glycogen** are all insoluble in water and are not sweet-tasting. In the bodies of both plants and animals polysaccharides are used as a convenient way to store simple sugars. Glycogen is found in animal cells and starch in plant cells. Both can be converted to simple sugars by hydrolysis.

Carbohydrates provide the main source of energy in the diet, and are therefore the main ingredient for cell respiration. Excess simple sugars can be converted into glycogen and stored in the liver and muscles until needed. Excess sugars can also be converted into fats and stored beneath the skin.

3. Fats. Fats, like carbohydrates, contain only hydrogen, oxygen and carbon. They are complex organic substances comprising one molecule of **glycerol** joined to three molecules of **fatty acid.** Fats differ from each other according to their constituent fatty acid or acids. Thus tristearin is a fat made up of one molecule of glycerol joined to three molecules of stearic acid and in tripalmitin the three fatty acids are each palmitic acid.

An **oil** is a fat with a low melting point, which means that it is liquid at room temperature. Together, fats and oils, as well as waxes, are called **lipids.** Lipids are an essential ingredient of the human diet because, like glycogen, they can act as a food store. Excess food substances can be converted into lipids and laid down under the skin where they can insulate the body from heat loss. They can also be found in large masses cushioning many of the body's vital organs, such as the kidneys. Lipids in combination with proteins are also a structural component of cell membranes and cell organelles, such as mitochondria.

Although some lipids can be synthesized from carbohydrates, many cannot. Lipids therefore have to be included in the diet (see page 40).

4. Nucleic acids. Two types of nucleic acid are found in cells. These are deoxyribonucleic acid (DNA) and ribonucleic acid (RNA). DNA is found only in the nucleus of the cell. RNA is mainly found in the cytoplasm. Nucleic acids are long-chain molecules (like proteins) made up of sub-units called nucleotides. They are much larger molecules than proteins and the nucleotides are more complex than amino acids.

Nucleotides consist of three molecules linked together. These molecules are phosphoric acid, an organic base and a sugar. In DNA this sugar is deoxyribose and in RNA it is ribose. The organic bases are adenine, guanine, cytosine, thymine (in DNA) and uracil (in RNA). The order in which these bases are linked up in the nucleic acid chain forms a code (the genetic code) which determines the characteristics of the cell (see page 167).

5. Inorganic substances (a) Water. This ubiquitous substance provides the medium in which all biochemical reactions take place. Its importance as a medium for life is due to four of its chemical and physical properties; its ability to act as a solvent, its heat capacity, its surface tension and its freezing point.

If you drop crystals of sodium chloride (salt) into water they dissolve. This is because the water molecules weaken the association between the sodium and the chlorine atoms within the salt. They therefore part company and go into solution. Water is a good **solvent** because many substances will dissolve in it.

Heat capacity is the amount of heat required to raise 1 g of a substance by 1°C. Water has a high heat capacity compared to other liquids. This means that a large increase in heat results in a small increase in temperature of the water. Water is therefore good at maintaining a fairly constant temperature irrespective of wide fluctuations in the surrounding temperatures. Even on a very hot day the temperature of the sea remains fairly constant. This is very important to the organisms that live in the sea and to us, whose bodies are made up of about 70 per cent water.

Surface tension is the force that makes the surface of a liquid contract so that it occupies the least possible area. Water has the highest surface tension of any known liquid, and this is of great biological significance.

At low temperatures most liquids decrease in volume and increase in density. When such a liquid freezes the molecules pack densely together and the resulting solid sinks. The reverse happens with water. At its **freezing point** (0°C) its volume increases and its density decreases. Ice tends to float rather than sink. When the sea freezes, as in the polar regions, the coldest water is on the surface. The less cold water beneath is therefore able to sustain life.

(b) Acids. An acid is a compound that dissociates in solution to produce hydrogen (H^+) ions. Its strength as an acid is determined by the degree to which it dissociates in water. The acidity of a solution is called its pH. A pH of 7.0 is neutral. Acids have pHs below 7, alkalis or bases have a pH above 7. (The pH of pure water is 7.0.)

An example of a strong acid is hydrochloric acid (HCl). This acid is found in the stomach (see page 51).

(c) Bases and salts. A base is a compound that can combine with the hydrogen ions liberated in solution by an acid. Sodium bicarbonate is a base that is commonly found in the body.

Mineral salts are compounds of a metal with a non-metal substance. Sodium chloride (or table salt) is an example (see page 43).

Unicells and multicells

Some animal and plants cells are able to live independently of other cells. They are called unicells. Some examples are shown below (Figure 3).

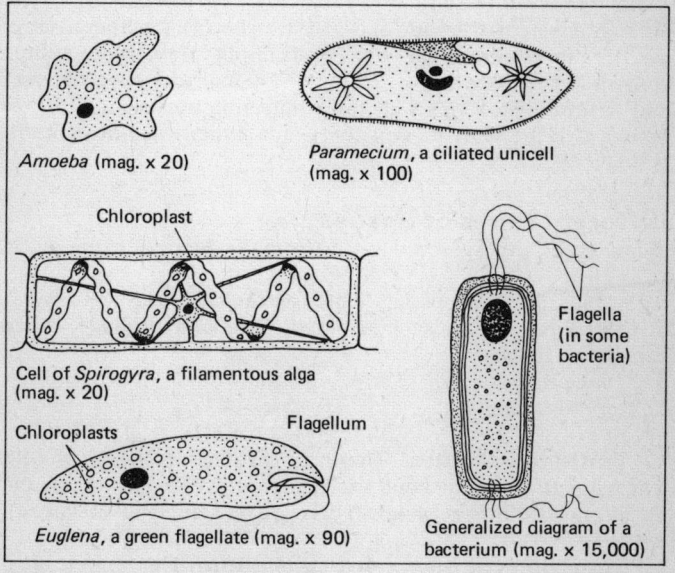

Figure 3. A variety of unicells

Most unicells are micro-organisms, which means that they cannot be seen by the naked eye. One of the most important group of unicellular micro-organisms to man is the bacteria, as many of them cause disease (see Chapter 9).

Most living cells exist as groups within one organism, which is known as a multicell. Man is an example of a multicellular organism. In the human body, cells are grouped together to form **tissues.** Tissues are grouped to form **organs,** and organs often work in groups called systems.

A tissue is a group of cells, each with a similar structure and a similar function to perform. An example is muscle.

An organ, such as the kidney, is a collection of different tissues which are grouped together to perform specific vital functions. The kidney consists of epithelial tissue, connective tissue, blood and muscle, which together form an efficient system to filter the blood of its waste products and to regulate salt and water levels in the body.

An organ system is a group of organs which work together to carry out the same overall function. For example, the nervous system consists of the following organs: brain, spinal cord, eye, ear, nose and tongue. These, and all the connecting nervous tissue, together give the body its sensitivity and ability to co-ordinate its actions.

Different types of tissues
Human cells are organized into **four** major types of tissue.

1. Surface and lining tissues, called epithelial tissues.
2. Connecting tissues such as blood, bone and cartilage, called connective tissues.
3. Muscle tissues.
4. Nervous tissues.

1. Epithelial tissues. These are protective sheets of cells which line the inner and outer surfaces of the body, its organs and the internal surfaces of glands. There are six main types (Figure 4):

(a) Squamous or pavement epithelium. This is made up of very flat delicate cells and is found wherever a protective lining needs to be highly permeable to molecules in solution. For example, the lining of blood capillaries and the inner surfaces of the lung alveoli (see page 64).

(b) Columnar epithelium. Cells of this type of tissue are more substantial and get their name from their column-like shape. Columnar epithelium is found in the larger ducts of the kidney, the gall bladder, stomach and small intestine. The inner surface of the cells is often greatly folded, forming structures called microvilli. Microvilli increase the surface area of each cell to aid the absorption of particles.

(c) Ciliated epithelium. Here the cells resemble in size those of columnar epithelium. In addition they bear numerous small hairs, or cilia, on their outer surface. The cilia are able to beat rapidly and rhythmically. This tissue is found wherever fluids and particles need to be moved along a tube. For example, it lines the surface of the lungs, where the cilia trap and expel dust and other foreign materials. Cigarette smoke destroys the cilia of the lung epithelial tissue.

(d) Cubical or cuboidal epithelium. The least specialized form of epithelium; the cells are cubical in shape. It is found lining various glands in the body, e.g. the thyroid gland in the neck.

(e) Glandular epithelium. This tissue contains both columnar epithelial and secretory cells. It forms the lining of the small intestine, where it produces mucus to aid the flow of the food.

(f) Stratified epithelium. Unlike the other epithelia, this tissue is multi-layered. It is found wherever tissue is susceptible to wear through friction, for example, the skin and the linings of the mouth and oesphagus.

2. Connective tissues. The function of connective tissue is support. It binds together the organs and other tissues of the body. Connective tissue is therefore very strong. All connective tissues consist of a ground substance or matrix (like a sort of glue) in which a variety of structures and cells are embedded. There are five types of connective tissue:

(a) Areolar connective tissue. This fills up the spaces between organs. Within the matrix are two types of fibre and four types of cell. The large fibroblast cells produce the fibres – unbranched collagen fibres which run parallel to one another in bundles, and branched elastic fibres which criss-cross throughout the matrix. The mast cells secrete the matrix, the fat cells are stores of fat and the macrophages can digest foreign particles.

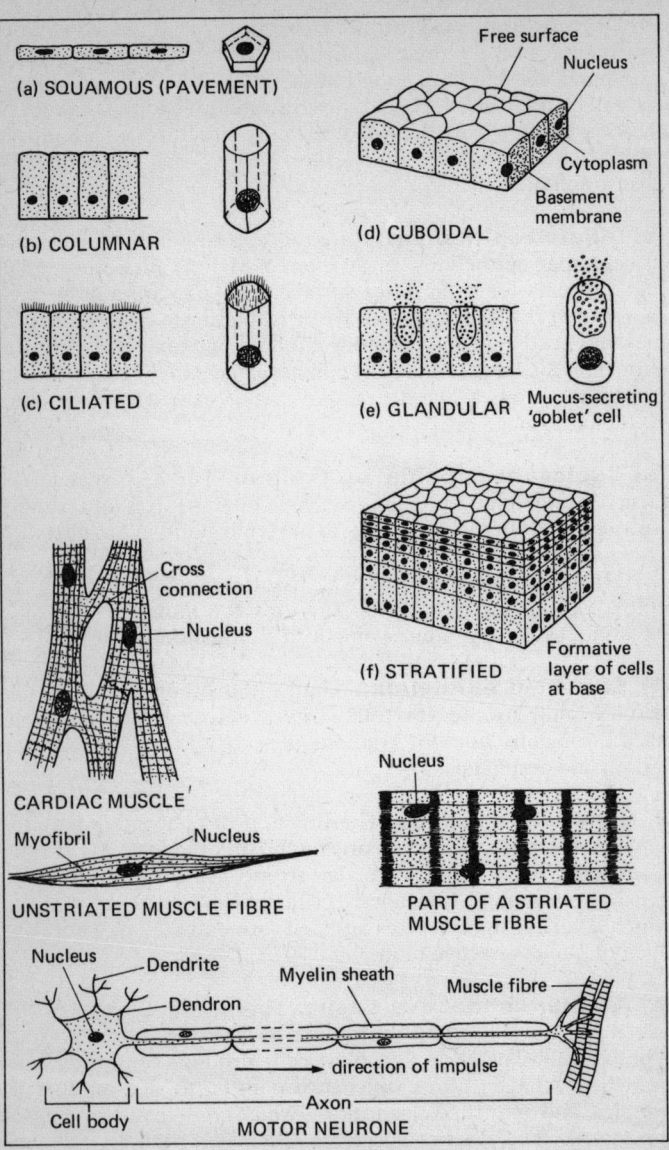

Figure 4. Different types of epithelial, nervous and muscle tissue

Areolar tissue can be found all over the body with varying numbers of the different cells and fibres. For example, in the adipose tissue beneath the skin the matrix contains fat cells and little else.

(b) Fibrous connective tissue. This is areolar tissue consisting almost entirely of collagen fibres. It is inelastic and very tough. It forms tendons, which attach muscles to bones. The Achilles tendon, for example, connects the calf muscle to the heel bone.

(c) Elastic connective tissue. Here, instead of collagen fibres, elastic fibres predominate. It is much more elastic than fibrous connective tissue and is found in ligaments, which hold joints together.

(d) Skeletal connective tissue. This supports the body and provides it with a rigid framework. There are two kinds: **cartilage** and **bone.** Again both consist of cells embedded in a matrix, but this time the matrix is rigid and hard.

Cartilage is made up of cells called chondroblasts that are embedded in a blue-coloured matrix, chondrin. The chondroblasts secrete the chondrin. The simplest type of cartilage consists of just these two components. It is known as hyaline cartilage and is found as rings lining the trachea (or windpipe). The rings help the trachea maintain its cylindrical shape and ensure that it remains open for the passage of air into the lungs.

Other forms of cartilage contain collagen or elastic fibres. Fibrous cartilage (containing collagen) is found in the intervertebral discs between the bones of the spinal column. Elastic cartilage is found in the pinna of the ear (ear lobe) and in the nasal septum of the nose.

Bone is much harder than cartilage. This is because its matrix contains calcium salts, chiefly calcium phosphate. These salts are secreted by special cells, osteoblasts, that are arranged in concentric circles round nerves and blood vessels running the length of a bone. Despite its rigidity, bone is very much a living tissue. A system of channels connect the osteoblasts to the blood vessels, so that all the cells' needs can be met.

(e) Blood. A type of connective tissue that is almost always on the move. It consists of three types of cell suspended in a fluid matrix (or medium) called plasma. This is dealt with in detail in Chapter 4.

3. Muscle tissues. This group of tissues comprises specialized cells that are able to shorten their length and then return to their usual size. The cells are referred to as contractile fibres. In a muscle, the contractile fibres are bound together by connective tissue (see Figure 4).

When a muscle contracts the protein molecules within each of the fibres rearrange themselves. This process requires energy.

There are three types of muscle tissue:

(a) Voluntary or skeletal muscle. This makes up the muscles that move the bones of the body, hence one of its names. Each muscle consists of many bundles of fibres within a connective tissue sheath. The muscles are also called voluntary because they are under control of the conscious part of the brain. This type of muscle tissue is sometimes called striped or striated muscle because the fibres appear striped under a light microscope. Voluntary muscle fibres can be up to 40 mm in length, but when contracted can be considerably shorter.

(b) Involuntary or smooth muscle. A type of muscle found in the walls of structures such as blood vessels, the small intestine and the uterus. The tissue cannot usually be controlled by conscious thought therefore cannot be contracted at will. Each muscle fibre is a single cell which is elongated and pointed at each end. The fibres can contract more slowly than those of skeletal muscle and can remain contracted for longer periods of time. They are found in circular or longitudinal layers around many of the body's tubes or ducts. In the gut their contraction pushes food along, the process of peristalsis (see page 51)

(c) Cardiac muscle. This type of specialized muscle is found only in the heart. It contains single cells that are branched and joined together (see Figure 4). When one cell contracts the whole cardiac muscle contracts. These contractions can be maintained at a steady rate for the whole of a person's life.

4. Nervous tissue. A tissue containing cells called neurones. Each neurone (see Figure 4) consists of a cell body and a number of protruding processes, axons and dendrons, which are able to conduct small electrical impulses.

Axons and dendrons are usually surrounded by an insulating fatty sheath, the myelin sheath, which speeds up the passage of an electrical impulse along the fibres. There are three types of neurone: motor, sensory and connector (see Chapter 6).

Processes taking place within the cell

Living cells are very different from dead ones. Inside living cells there is much going on – many thousands of chemical reactions which are carefully controlled so that the cell continues to function normally. Dead cells, by comparison, can only rot away.

The chemical activity that takes place within the body's cells is known as **metabolism** and the products of this metabolism as metabolites. Even cells of tissues such as bone, cartilage and blood are metabolically active. It is only the body's dead structures, such as finger nails and hair, that do not metabolize.

Metabolic reactions are either **anabolic** or **catabolic.** Anabolic reactions involve the building up of large molecules from smaller ones, i.e. $A + B = AB$. A and B are the reactants and AB is the product of the reaction. An example would be the build up (synthesis) of the polysaccharide glycogen from simple sugars (see page 14).

Catabolic reactions involve the breaking down of large molecules into smaller ones, i.e. $AB \rightarrow A + B$. Here AB is the reactant and A and B are the products. An example is the breakdown of glycogen in the liver to form its constituent simple sugars.

The most important difference between the two types of reaction is that anabolic reactions use up energy and catabolic reactions tend to liberate energy. This point will be returned to shortly when considering cell respiration.

Metabolic reactions are different from the chemical reactions that you may have seen taking place in a chemistry laboratory. For instance, you may have thrown sugar (sucrose) into a bunsen burner flame and seen it burn vigorously. Inside a cell, sugars are also burnt (oxidized), but never so vigorously that they ignite. Instead they are oxidized gradually by a series of reactions, each reaction being assisted by a different catalyst or enzyme. All metabolic reactions are like this. They proceed slowly along a pathway of gentle reactions.

Enzymes

Enzymes are organic catalysts that speed up chemical reactions within living cells. Without enzymes these reactions would take place very slowly, or not at all. As well as speeding up reactions, enzymes also control the order in which they take place. This is achieved because each reaction is catalysed by a specific enzyme and not by any other.

Enzymes are proteins. Their properties can be summarized as follows:

1. Enzymes work very rapidly.
2. Enzymes are not destroyed by the reactions they catalyse.
3. An enzyme will catalyse a reaction in either direction. For example, the synthesis or breakdown of glycogen requires the same enzyme.
4. Enzymes are inactivated by high temperatures (see page 25).
5. Enzymes are affected by the acidity (pH) of the reactants (see page 25).
6. Enzymes are reaction-specific. Normally one enzyme will catalyze only one reaction or type of reaction.

Some substances will inhibit or interfere with the action of enzymes. Inhibitors are therefore metabolic poisons because they slow up or stop metabolic activity. Arsenic, cyanide and strychnine are examples of metabolic poisons.

The majority of enzymes work inside the cell but some, such as the digestive enzymes, work outside. The reason for this will be apparent in Chapter 2.

Experiments with enzymes

1. Effect of temperature on enzyme action. Diastase is an enzyme that will hydrolyze starch into maltose. Place 5 mm^3 of diastase into each of five test-tubes labelled A, B, C, D and E. Place tube B in a water-bath at 25°C, tube C in a water-bath at 40°C, tube D at 60°C and tube E at 100°C. Tube A should be left at room temperature, which should be measured. Keep the tubes in their respective places for 5 minutes. At the end of this period, add 5 mm^3 of fresh starch solution to each tube and thoroughly mix. Take a note of the time at which this is done. Then at 5-minute intervals test each tube for the presence of starch. This is done by taking a few drops of the incubating mixture and mixing it with one or two drops of iodine on a white tile. If starch is still present the iodine will turn blue-black. Continue the experiment for at least an hour. Note how long it takes for starch to disappear from each tube. Then plot your results on a graph like the one below.

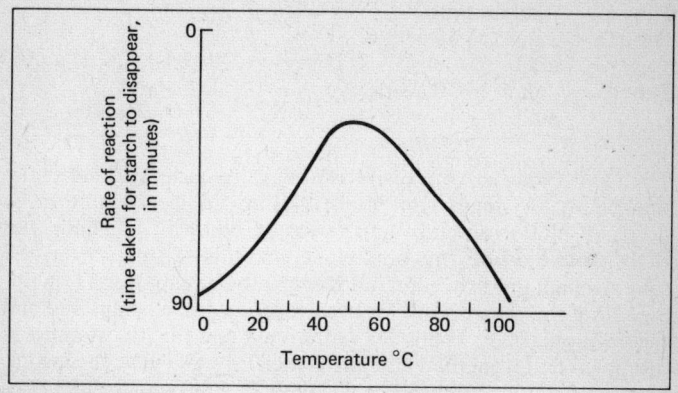

Figure 5. Effect of temperature on enzyme action

You should find that the reaction proceeds quickest at about 35°C.

Figure 5 shows the effect of temperature on the activity of most human enzymes. Up to 40°C an increase in temperature produces an increase in the reaction rate. Above 40°C the reaction rate slows down as the temperature is increased. This is because at high temperatures the enzyme has been denatured. In other words, its structure has been altered by the heat and it is no longer able to catalyse the reaction. The reaction works best at about 40°C, which is known as the **optimum** temperature. This roughly corresponds to the internal temperature of the human body (37°C).

2. Effect of pH on enzyme action. Amylase is an enzyme that is found in human saliva. It converts starch into maltose. The effect of pH on the activity of amylase can be determined by a similar experiment to number 1 above. Place 5 mm^3 of starch solution into each of five test-tubes labelled A, B, C, D and E. Then add to each tube in turn acid or alkali according to the table overleaf:

Tube	Acid/Alkali	Approx. pH	Reaction time
Tube A	1 mm^3 M/20 sodium carbonate solution	9	
Tube B	0.5 mm^3 M/20 sodium carbonate solution	7½	
Tube C	Add nothing	7	
Tube D	2 mm^3 M/10 acetic acid	6	
Tube E	4 mm^3 M/10 acetic acid	3	

Then add 1 mm^3 of your own fresh saliva to each tube and shake thoroughly. At regular intervals take up from each tube three drops of solution and mix with a drop of iodine on a ceramic tile. If the iodine drop turns blue-black starch is still present in the tube. Repeat this procedure until each tube has no starch remaining. Note the time it took for each reaction to be completed and record it on a chart as above. You should find that the starch will disappear first from tube C. Tube B then tube A will be the next to lose their starch, followed by D, then E. Plot your results on a graph similar to the one below. The optimum pH for amylase is about 7. Acid conditions drastically slow the reaction. Alkaline conditions slow the reaction, but to a lesser extent.

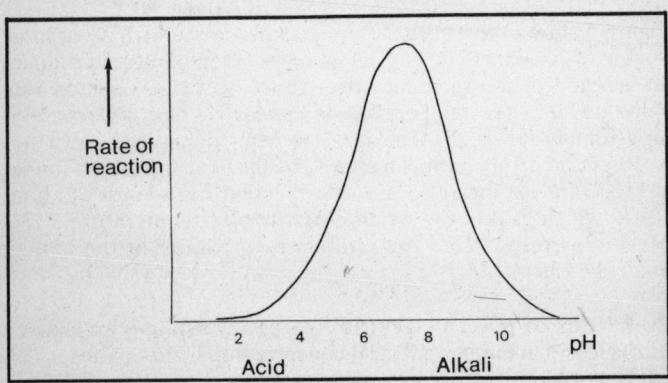

Figure 6. Effect of pH on enzyme action

Enzymes are very sensitive to the pH of their surroundings, and most enzymes can function only within a narrow range of pH.

However the optimum pH for any enzyme varies considerably. For example, salivary amylase (Figure 6) has an optimum pH of just over 7 (slightly alkaline); rennin, which works in the stomach, has an optimum pH of about 2 (very acid).

Movement of substances in and out of the cell

Substances pass through the plasma membrane by four main processes: diffusion, osmosis, active transport and phagocytosis.

1. Diffusion. If you drop a crystal of copper sulphate into a container of water, the blue colour of the crystal gradually spreads throughout the water. This is because the copper sulphate molecules move away from the crystal, where they are concentrated, to where there are fewer copper sulphate molecules, i.e. in the water. In time the crystal dissolves completely, distributing the copper sulphate evenly throughout the water.

This process is known as diffusion. It involves the movement of molecules from a region of high concentration to a region of low concentration. (The same principle applies to the way people tend to spread themselves out on a beach or on a bus.)

Diffusion is a means by which gases or liquids move in and out of cells. Consider the example of oxygen. Oxygen is continually being used up in a cell, so a cell will tend to have a lower concentration of oxygen than the surrounding tissue fluid. Oxygen will therefore diffuse continuously from the tissue fluid to the cell. The barrier to diffusion is the plasma membrane. For a substance to diffuse through it, it must be permeable to that substance. Oxygen and carbon dioxide, being small-sized molecules, readily pass through. Large molecules such as proteins cannot.

2. Osmosis. This is the movement of water from a weak solution to a strong solution through a semi-permeable membrane. A weak solution is really a high concentration of water molecules with a small concentration of dissolved solute molecules. A strong solution contains less water and more solute. This means that osmosis is effectively a form of diffusion. The water molecules move from where they are plentiful to a region where they are scarce.

The plasma membrane is permeable to water, so water molecules will pass in and out of a cell according to the strength of fluids within and external to it.

The plasma membrane is impermeable to most dissolved substances, such as glucose, or to molecules in suspension, such as proteins.

The **osmotic pressure** of a cell is the force exerted by the cytoplasm to draw in water from the outside. If the blood and tissue fluid become too watery, water will be drawn into general body cells by osmosis. By the same token, if the blood and tissue fluid contain little water, their osmotic pressure will draw water out of the cells. Wild fluctuations in the water content of the cells can seriously affect their metabolism, so the osmotic pressure of the blood has to be carefully controlled. This process of osmoregulation is discussed in Chapter 5.

3. Active transport. Diffusion and osmosis do not require energy, molecules simply move down a concentration gradient until an equilibrium situation is reached. Active transport, as its name suggests, is an active, energy-requiring process (Figure 7).

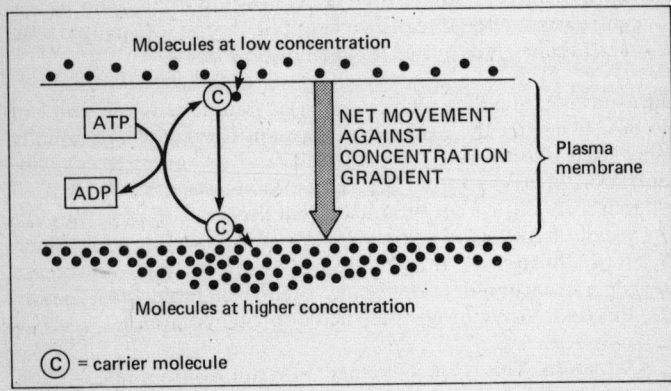

Figure 7. Active transport

It takes place wherever substances are moved across the plasma membrane against a concentration gradient. That is, moved from an area of low concentration to an area of high concentration. Energy is required to do this because the molecules are being moved against their tendency to move in the opposite direction. The exact mechanism for active transport is not known, but one theory is that the molecules are moved across the membrane by a carrier molecule. An example is the uptake of iodine by cells in the thyroid gland (See page 118).

4. Phagocytosis. Literally, this term means 'cell-eating'. The process was first seen in white blood cells (see page 74), which use the process to engulf bacteria in order to destroy them.

Cell respiration

Cell respiration is the catabolic breakdown of organic compounds, chiefly carbohydrates, to produce energy. Oxygen is usually required. Carbon dioxide and water are produced as waste products. This process is sometimes called tissue or internal respiration (to distinguish it from external respiration or, more commonly, breathing). It occurs in **all** living cells.

Cells that require oxygen to respire are referred to as **aerobes** and those able to respire in the absence of oxygen are called **anaerobes.**

Aerobic respiration is usually represented as a single chemical reaction:

$$C_6H_{12}O_6 + 6O_2 = 6H_2O + 6CO_2 + 2,880\,\text{kJ energy}$$
$$\text{glucose} \quad \text{oxygen} \quad \text{water} \quad \text{carbon dioxide}$$

but this is very misleading because many separate reactions are involved, not one. It is however useful as an overall summary of the process of respiration.

The energy obtained from respiration originates in the chemical bonds of the glucose molecule, and as this molecule is gradually broken down, so small amounts of energy are released. However, this energy may not be immediately required for the cell's functioning, so it must be stored in some way. This is achieved by using the energy to convert the chemical adenosine di-phosphate (ADP) to **adenosine tri-phosphate** (ATP). The conversion is anabolic as it converts two reactants, ADP and phosphate, into one product, ATP. Energy is required for this reaction and is taken up to form the additional phosphate bond of the ATP molecule. Later, when energy is required, the ATP readily breaks down into ADP and phosphate, releasing the energy of this bond.

The catabolism of 1 glucose molecule eventually leads to the synthesis of 32 molecules of ATP.

In the initial stages of respiration, the 6-C atom glucose molecule is broken down into two identical 3-C atom molecules, pyruvic acid. This process is called **glycolysis** and occurs in the cytoplasm of the cell. A small amount of ATP is produced. Next, in the lumen of the cell's mitochondria, each pyruvic acid molecule is broken down in the presence of oxygen through a cycle of reactions that yield the bulk of the glucose molecule's energy in the form of ATP molecules. Heat is also generated. The carbon dioxide and water produced diffuse out of the mitochondria, through the plasma membrane and into the bloodstream.

This second stage of respiration is known as the **Kreb's cycle,** named after the biochemist who discovered it some fifty years ago.

In the absence of oxygen, pyruvic acid is converted to lactic acid (see Figure 8), which is stored until sufficient oxygen is available to oxidize it. Thus the formation of lactic acid from glucose represents anaerobic respiration. It takes place in human muscle cells during vigorous exercise, when oxygen may be in short supply. Some ATP can be produced in this way.

Man is therefore an example of a partial anaerobe; he will resort to anaerobic respiration should it become necessary. Some organisms, such as yeasts and some bacteria, are complete anaerobes in that they are totally independent of oxygen. Indeed, in some cases they may be poisoned by oxygen, even in low concentrations.

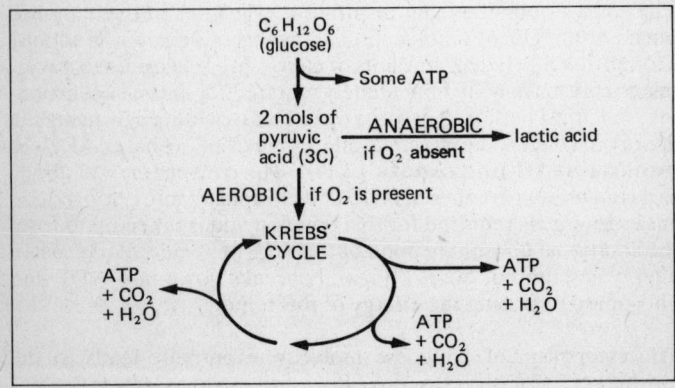

Figure 8. Summary of cell respiration

The anaerobic respiration of yeast cells is of great significance to man because instead of converting glucose to pyruvic acid and then to lactic acid, the pyruvic acid is converted to alcohol.

$$C_6H_{12}O_6 \rightarrow 2C_2H_5OH + 2CO_2 + 210 \text{ kJ energy}$$
$$\text{glucose} \qquad \text{alcohol}$$

This reaction is known as **fermentation** and is the basis of all wine, beer and spirit manufacture.

Cell multiplication

Cells multiply in number to produce an increase in body size and/or for tissue repair. The type of cell division involved is called **mitosis.** Mitosis occurs most often in the early stages of growth and development, when the tissues and organs are forming. Later as cells become more specialized, this power of division is lost. In adults, few cells retain the ability to divide and multiply. One area of the body where mitosis is particularly common throughout life is the skin.

When cells undergo mitosis, the same basic pattern is followed. First the nucleus divides into two and then the whole cell divides, forming two identical cells (see Figure 9). Both daughter cells then enlarge to the same size as the original cell.

Human cells, with the exception of sperm and ova, have 46 chromosomes. A single chromosome is shown in Figure 10. At one point along each chromosome is a granular structure, the centromere. Chromosomes vary in length, but for each size of chromosome in the nucleus there is one other of equal length. In other words, the 46 chromosomes can be grouped as 23 pairs.

When chromosomes first become visible during mitosis they appear to be double-stranded. This is because just previously each chromosome has replicated itself to form two identical partners or chromatids. These are joined at the centromere. The cell is now ready to divide.

The following stages of mitosis are distinguishable:

(a) Interphase. During this period, the resting stage, chromosomes occupy the nucleus in the form of fine threads but they are not separately visible. The dark mass near the centre of the nucleus is the nucleolus. A pair of dots visible in the cytoplasm represents the divided centriole.

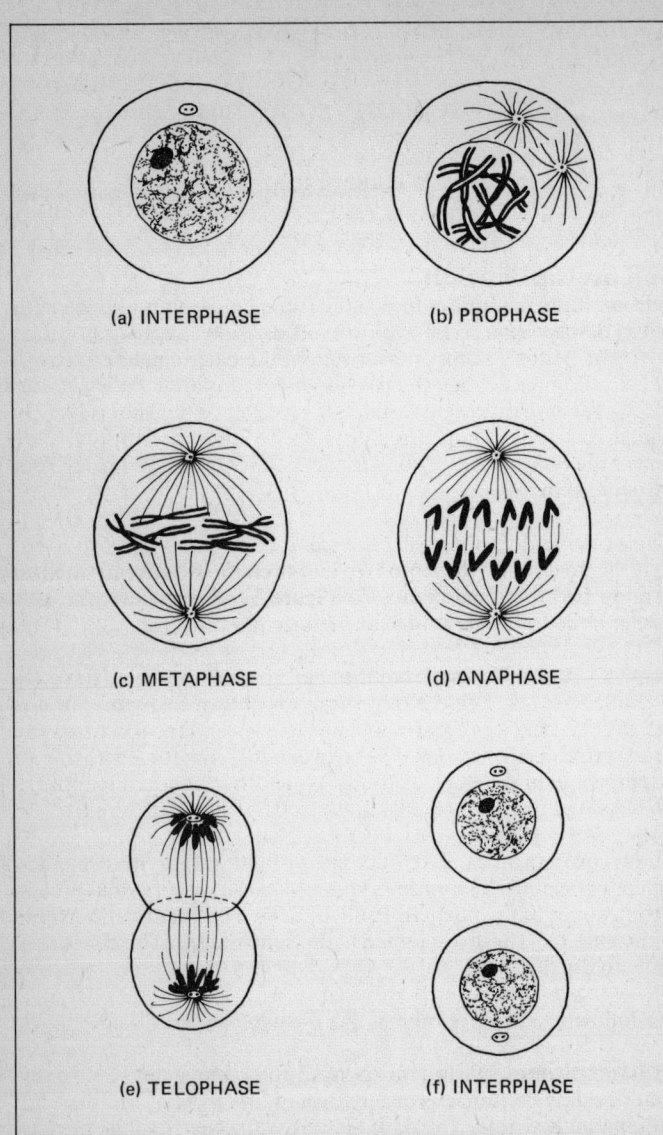

Figure 9. Diagram of mitotic stages

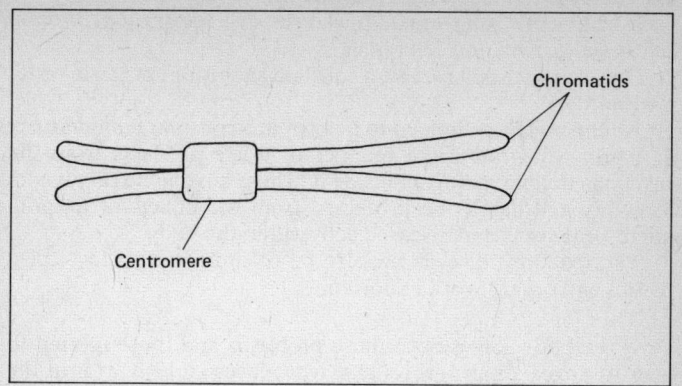

Figure 10. Appearance of chromosomes at anaphase

(b) Prophase. Chromosomes are seen in contracted but still elongated form. The centrioles begin to separate and to make a spindle.
(c) Metaphase. The chromosomes are best studied at this stage. They lie in a horizontal plane; the nuclear membrane has disappeared.
(d) Anaphase. On each chromatid pair (a chromosome), the centromeres separate and are drawn towards the poles of the cell.
(e) Telophase. Two cell membranes form between the poles.
(f) Interphase. Nuclear membranes have been reconstructed and the two daughter cells enter the resting stage.

The result of mitosis is two identical cells, each with an identical complement of 46 chromosomes.

Sperm and ova are unusual in that they each possess only 23 chromosomes (one of each of the 23 pairs). They are produced not by mitosis, but by a different process of cell division, meiosis, described on page 151.

Man as a multicellular organism
Man has a large multicellular body that is made up of a number of distinct internal organs and specialized tissues. Because of this the majority of human cells are situated far from the surface of the body. This poses problems for both the cells and organism as a whole.

1. How can each cell receive a constant supply of all the nutrients,

such as glucose, that are required for cell respiration, protein synthesis, growth and cell repair?
2. How can each cell receive a constant supply of oxygen in order to respire?
3. All the body's cells need to be kept at a constant temperature.
4. Every cell produces a number of waste products from the reactions that are going on inside it. These have to be removed.
5. Every cell has to be protected from the effects of harmful micro-organisms and diseased cells within the body.
6. The reactions of cells need to be co-ordinated so that cells, tissues and organs work in unison.

How the body solves both these problems and those related to how it moves about, grows and reproduces is dealt with in the remaining chapters of this book. But first an important concept is introduced which will recur throughout these topics and is fundamental to an understanding of why the human body works in the manner it does.

The external and internal environment. Figure 11 is a schematic representation of the human body in section. The **body surface** is the boundary between the body and its surroundings, the **external environment.** Everything that lies within the body surface is part of the organism and is referred to as its **internal environment.**

The body surface has two major functions. Firstly, it protects the internal environment from the constantly changing and challenging nature of the external environment. Secondly it serves to retain all the contents of the body and keep the body intact. At various points the body surface has become pushed in on itself, as if a fist has punched hollows into it. The hollows are called **invaginations.** Notice though that the body surface remains intact at these points.

These invaginations are important because they are special areas of the body surface where vital substances such as oxygen and food can be exchanged between the external and internal environment. For instance, the structure marked A represents the gut, which is a hollow tube running through the body. The gut is bounded on all sides by the body surface so that its interior is really an area of external environment and not part of the body at all. As you will see in Chapter 2, the body surface lining the gut has become specialized to allow food material to pass through it.

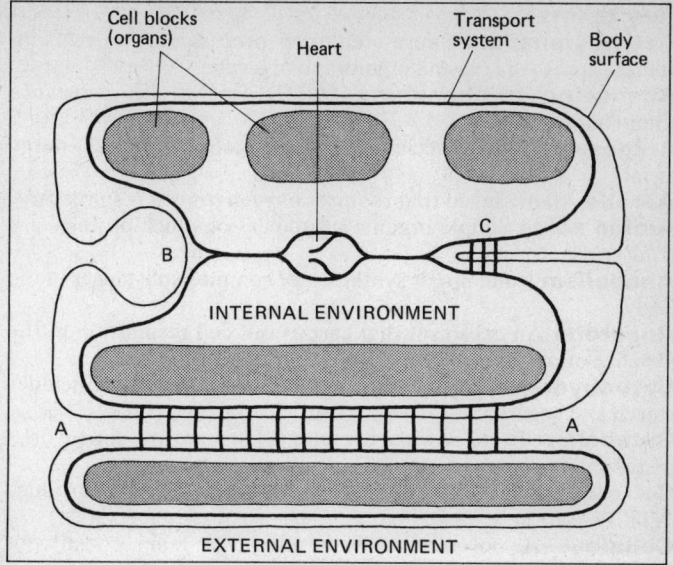

Figure 11. Schematic representation of the human body

Invagination B represents the lungs. Here again the body surface encloses an area of external environment but as the lining of the lungs it is specialized to allow the exchange of gases. This is described in detail in Chapter 3.

Structure C represents a Bowman's capsule, which is part of the kidneys. Here waste products pass into the external environment of the bladder, from where they leave the body. This process will be dealt with in Chapter 5.

The other structure in Figure 11, a central chamber and a system of solid lines, represents the body's transport system. For the same reason that people need buses and trains in order to get to work, the body needs an efficient transport system to carry oxygen and nutrients from the external environment to the body's cells and to carry waste products in the opposite direction. The human body's transport system is made up of a series of tubes containing a special fluid (the blood) and a pump (the heart), which makes the fluid flow. Together they comprise the vascular system and you will find out more about this in Chapter 4.

Key terms
Active transport Energy-requiring process that moves substances across the plasma membrane of a cell.
Adenosine tri-phosphate (ATP) Energy-rich substance found inside cells.
Adipose tissue A specialized group of cells that stores energy as fat.
Aerobe An organism that requires oxygen for cell respiration.
Amino acids Simple organic substances of which proteins are composed.
Anabolism Build up, or synthesis, of complex substances in the body.
Anaerobe An organism that carries out cell respiration in the absence of oxygen.
Carbohydrates Large class of organic compounds that includes starch and sugars.
Catabolism Breakdown of complex substances into simple ones inside the body.
Cell respiration Breakdown of sugars inside the cell to produce ATP.
Cellulose A polysaccharide forming the main constituent of the wall of plant cells.
Centriole Cell organelle important in cell division.
Chlorophyll Light-absorbing green pigment found only in plants.
Chloroplasts Plant cell organelles, containing chlorophyll; site of photosynthesis.
Chromosome Thread-like structure found inside the cell nucleus; carries blueprint for cell's activity.
Cristae Finger-like processes of the inner wall of a mitochondrion.
Cytoplasm That part of the protoplasm outside the nucleus of a cell.
Deoxyribonucleic acid (DNA) Large molecule of which chromosomes are composed.
Diffusion Movement of molecules from a high concentration to a low concentration across a semi-permeable membrane until the concentrations are equal.
Diploid Having a full set of paired chromosomes; in man 46 chromosomes (23 pairs).
Endoplasmic reticulum Part of a cell where proteins are processed for 'export'.
Enzymes Substances found inside cells that speed up chemical reactions; all are proteins.

Epithelium A tissue that lines the surface of the different parts of the body.
External environment Everything that is outside the body surface of an organism.
Fat An organic compound comprising fatty acids and glycerol; also known as lipid.
Golgi body Cell organelle that is thought to store proteins prior to 'export'.
Glucose A simple sugar; broken down to water and carbon dioxide during cell respiration.
Haploid Having a set of unpaired chromosomes; in man 23 chromosomes.
Inorganic substances Substances which do not contain carbon but including carbon dioxide.
Internal environment The part of the body retained by the body surface.
Lysosome Sac-like cell organelle containing destructive enzymes.
Meiosis Cell division in which the chromosome number becomes halved.
Metabolism The chemical reactions going on within cells.
Mitochondrion Organelle present in all cells; site of cell respiration.
Mitosis Cell division producing two genetically identical cells.
Multicell An organism comprised of more than one cell.
Nucleolus Dense area within a nucleus; site of production of RNA.
Nucleotide Building block of a nucleic acid such as DNA.
Nucleus Organelle that contains the chromosomes and thus controls a cell's activity.
Organic substances Chemicals that contain carbon excepting carbon dioxide.
Osmosis Movement of water from a weak solution to a strong solution through a semi-permeable membrane.
Organelles Structures found inside a cell.
Peptide bond The link between two amino acids within a protein molecule.
Phagocytosis Process by which certain cells engulf or 'eat' bacteria.
Photosynthesis Process by which green plants manufacture sugars using energy from the sun.
Plasma membrane Living boundary of cells.
Polysaccharide Long-chained carbohydrate made of simple sugars, e.g. starch.

Protein Essential building block of protoplasm, made up of amino acids.

Protoplasm The living material of cells, consisting of cytoplasm and nucleus.

Ribonucleic acid (RNA) Similar to DNA but found largely in the cytoplasm.

Ribosomes Organelles composed of RNA; site of protein synthesis.

Semi-permeable membrane Allows small molecules such as water or oxygen to pass through more easily than larger molecules such as glucose.

Unicell A living organism consisting of one cell. e.g *Amoeba*.

Chapter 2
The Gut and Nutrition

Living organisms must obtain food from their external environment in order to survive. Food contains the simple inorganic and organic substances which are required for cell respiration, growth and repair and the synthesis of proteins. The process of obtaining food and making its constituent chemicals available to the body's cells is called **nutrition.**

Human food is complex in chemical structure and needs to be broken down considerably by the body before it can be utilized. In common with most animals, man possesses a long tube, the **gut** or alimentary canal, in order to do this. Beginning at the **mouth** and terminating at the **anus,** the human gut consists of a number of distinct regions or compartments, with associated organs, each having a sophisticated food-processing function (see Figure 12).

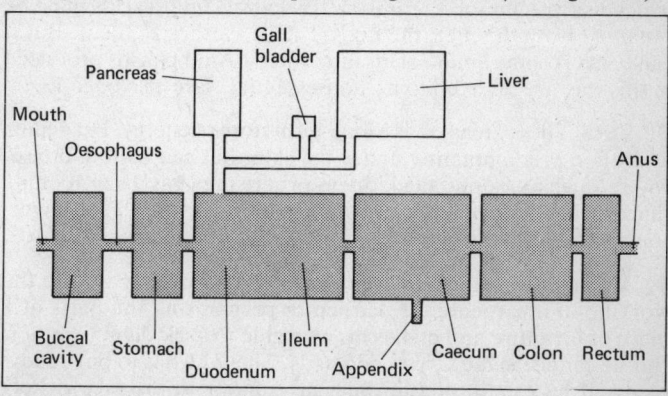

Figure 12. Schematic representation of the gut

The types and quantities of food eaten by an organism is called a **diet.** A typical human diet includes meat, dairy produce, fish, cereals, vegetables and fruits. A vegetarian diet excludes meat. A liquid diet, often given to hospital patients, contains only liquid food.

Food contains the following seven ingredients: proteins, fats, carbohydrates, water, vitamins, mineral salts and fibre. Some foods may contain just one of these ingredients. Refined white sugar, for example, consists almost entirely of carbohydrate. Other foods contain most or all of these ingredients. Consider a 100 g banana. This contains 1.1 g of protein, 19.2 g of carbohydrate and no fat. The remaining 79.7 g consists chiefly of water and fibre, together with vitamin C and trace amounts of minerals such as calcium. To remain healthy, a person's diet must include all of these ingredients in sufficient quantities.

Essentials of a balanced diet

1. Carbohydrates. These provide the body with a supply of energy. Excess carbohydrates can be stored as glycogen in the liver and muscles. Further excess supplies can be converted into fats and stored in the fat depots. For further details see page 117.

2. Proteins. A mixture of proteins is required for growth, cell replacement and the synthesis of enzymes and hormones. During digestion proteins are broken down into their constituent amino acids which are then absorbed into the bloodstream (see page 53). In the body's cells the amino acids are reassembled to produce whichever proteins are required. 'Essential' amino acids must be obtained from the diet; the body cannot synthesize them. But it can convert some amino acids into others. Amino acids produced in this way are referred to as 'non-essential'. See also page 13.

3. Fats. These are used as a long-term store of energy. Fat depots are found predominantly under the skin, but can be found elsewhere, such as around the kidneys, where they have a cushioning function. Fat depots are also important as an insulating layer, conserving the heat of the body. For more information see page 102.

4. Vitamins. The body's metabolism can be compared to the working of a carpenter. A carpenter prepares all the parts of a piece of furniture and makes his own glue to stick them together. But he cannot make screws and nails. These he has to buy ready-made. If he cannot obtain them he is likely to produce rickety furniture. The same applies to vitamins. You may survive without them, but not in good health, and not for very long. Vitamins are organic substances that metabolism cannot do without. The body cannot synthesize them so they must be provided in the diet. Only very small amounts are required, but if they are not obtained a typical deficiency disease results.

Scurvy is the best known of these diseases. It results from a lack of ascorbic acid (vitamin C). Scurvy was once a very common disease in Europe, but the introduction of root crops into the diet towards the end of the Middle Ages reduced its occurence unintentionally. Root crops and citrus fruits all contain ascorbic acid and their inclusion in the diet prevents the development of scurvy. However, the disease is still common among elderly people as they tend not to have a sufficiently varied diet. The symptoms include bleeding into the skin, around bones and joints and from the gums. The teeth become loose and misshapen. Resistance to infection is lowered. Lack of vitamin C produces a defect in the substance that binds cells together, particularly those in connective tissue and capillary blood vessels.

Vitamins are generally known by letters: A, B, C, D, etc, but it is more helpful to refer to them by their actual names as several of them are now known to be groups of many different substances. Vitamin B, for example, is a complex of at least twelve different chemicals. A full list of vitamins is given in Table 1.

Table 1. Vitamins and their characteristics

Name	Dietary source	Characteristics
Retinol (Vitamin A)	Fish oils, butter, meat, fresh vegetables, especially carrots.	Retinol forms part of one of the light-sensitive pigments of the eyes; it is essential for seeing in poor light. Its absence causes night-blindness.
Thiamine (Vitamin B_1)	Yeast, wheat-germ, brown flour and rice husks.	Vitamin B was originally thought to be one substance but now it is known to be composed of at least a dozen. Thiamine is involved in the use of oxygen in respiration by the tissues. Its absence gives rise to beri-beri, a wasting disease; this is common in Africa.
Riboflavin (Vitamin B_2)	Milk, meat and green vegetables	It is concerned with the utilization of energy from food by cells. Absence

Table 1. Vitamins and their characteristics (continued)

Name	Dietary source	Characteristics
		leads to symptoms similar to beri-beri.
Nicotinic acid (niacin)	Milk, meat and green vegetables.	It has functions similar to those of B_1 and B_2. Shortage of nicotinic acid leads to loss of weight, diarrhoea, skin and mental disorders, a condition known as pellagra.
Cobalamine (Vitamin B_{12})	Milk, meat and green vegetables.	Important for the synthesis of nucleic acids and therefore essential for the formation of new body cells, particularly red blood cells, which are being constantly replaced. Absence leads to anaemia.
Ascorbic acid (Vitamin C)	Citrus fruits, nuts and fresh vegetables.	Most plants and animals are able to synthesize vitamin C. Humans, apes, bats and guinea pigs cannot. Its absence in the human diet leads to scurvy (see page 40).
Calciferol (Vitamin D)	Cream, egg-yolk and cod liver oil. (It is also obtained from the action of sunlight on a substance in the skin. This is the most important source.)	Vitamin D is essential for the correct formation of human bone. It promotes the absorption of calcium from the intestine and assists in the movement of calcium ions while bone is growing. Its absence leads to rickets, which is abnormal bone formation leading to soft bones. Children with rickets have bow-legs.

Table 1. Vitamins and their characteristics (continued)

Name	Dietary source	Characteristics
Tocopherol (Vitamin E)	Wholemeal bread, butter.	Male rats deprived of vitamin E become sterile, and female rats cannot bear live young. There is little evidence that man ever lacks this vitamin.
Vitamin K	Vegetables, especially cabbage and spinach.	Vitamin K is required for the synthesis of the enzyme thrombin, without which blood is unable to clot (see page 77). The vitamin is also synthesized by bacteria in the human gut, from which it is absorbed. For this reason alone, its absence in adults is unlikely.

5. Water. This is an essential component of the human body, making up about 70 per cent of its weight. All of the body's fluids contain water and the enzymes that control metabolism can work only in solution. Water is also required to remove metabolic waste, forming urine, and to play a part in temperature regulation (see page 97).

6. Mineral salts. Like vitamins, mineral salts are required in only very small quantities, but if they are absent from the diet, ill-health results. Those such as sodium chloride (salt) exist in the body in an ionized form: $NaCl \leftrightarrows Na^+ + Cl^-$

A positively charged ion (Na^+) is called an anion, a negatively charged ion (Cl^-) a cation.

Table 2. Essential cations and anions

Name	Food source	Function in the body
Anions		
Iron (Fe^{2+})	Spinach, eggs, kidneys and liver.	Synthesis of haemoglobin for red blood cells. Absence leads to anaemia.

Table 2. Essential cations and anions (continued)

Name	Food source	Function in the body
Calcium (Ca^{2+})	Cheese, milk, beans, tap water in hard-water areas.	Essential for growth of teeth and bones, clotting of blood, muscle contraction, transmission of nerve impulses.
Sodium (Na^+)	Table salt.	Essential ionic component of blood plasma and other body fluids. Transport of carbon dioxide, nerve impulses.
Potassium (K^+)	Table salt, cereals.	Present in red blood cells and muscle cells. Similar functions to sodium.
Phosphorus (P^{3+})	Meat, eggs, and fish.	Synthesis of ATP and nucleic acids. Also needed for bone growth.
Cations Chloride (Cl^-)	Table salt.	Essential ionic component of plasma and other body fluids. Used to synthesize hydrochloric acid in the stomach.
Iodide (I^-)	Tap water, seaweed; often added to table salt.	Required for the synthesis of the hormone thyroxine, produced by the thyroid gland. Absence: goitre or cretinism.
Fluoride (F^-)	Tap water: exists naturally or is added.	Essential for the growth of teeth and helps to prevent tooth decay.

Most of these mineral ions are unlikely to be absent from the diet. Calcium, iron and iodine are the three most likely to be in short supply.

Very small quantities of cobalt, manganese, zinc and copper are also known to be essential, but a varied diet provides more than ample supplies. Cobalt is involved in the formation of vitamin B_{12} and zinc is involved in the synthesis of the hormone insulin.

7. Fibre. This consists of the indigestible components of plant material, which pass through the gut. The bulk of fibre is important because it stimulates the rhythmic contraction of the gut muscles, which propels the food along the colon and stimulates defaecation. One cause of constipation is insufficient fibre in the diet.

Food and energy

Energy comes in various forms: mechanical, electrical, chemical, etc. It may be stored, for example as in the spring of a clock, the cell of a battery or in the chemical bonds between atoms in molecules. One form can be converted into another: for example, in a motor electrical energy is converted into mechanical energy.

The human body requires energy to fuel its wide range of chemical reactions, to allow muscles to contract and to constantly heat its internal environment. This it obtains from the chemical energy stored within the bonds of simple sugars, fatty acids, glycerol and, to a lesser extent, amino acids, supplied by the diet.

The amount of energy contained in a sample of food is measured in **joules** (J) and kilojoules (kJ). A joule is thus a unit of energy and 1,000 joules = 1 kilojoule. In the past, this energy was measured in calories (cal), and people still refuse to eat chocolate "because it has too many calories". 1 calorie = 4.2 joules. A calorie is a unit of energy defined as the amount of heat required to raise the temperature of 1 g of water by 1°C.

The amount of energy released from food depends upon its chemical nature. Fats yield more than twice as much energy per gram than carbohydrates and proteins.
1 g of fat yields 39 kJ of energy.
1 g of carbohydrate or protein yields 17 kJ of energy.

The daily requirements of energy naturally depend upon the person concerned. Someone who carries out physical work on an oil rig requires much more energy than a person who sits at a desk and writes textbooks. Large energy requirements must be met by a large intake of food. Table 3 shows the range of daily energy requirements of different people. Energy requirements vary according to body size, age, sex, occupation, leisure activities and special conditions such as pregnancy.

The nature of work also determines the rate at which energy is used or the **metabolic rate.** The basal metabolic rate is that at which energy is used up when a person is in a complete state of rest, i.e. with only the essential homeostatic processes such as breathing going on. This basal rate is highest at birth and decreases slowly throughout life. Men have a higher basal rate than women, but rates vary considerably and are genetically determined (see page 167).

Table 3. The range of daily energy requirements

Age and occupation	Energy used in an average day in kJ	Special dietary needs
Birth to 1 year	0-6 months 2800 6-12 months 4000 (both sexes)	From 0-6 months 2 g of protein per kg of body weight are required. This amount reduces to 1 g by age 1
8 years Active child	Both sexes 8800	30 g of protein are needed per day. The DHSS recommend 53 g per day.
15 years Active child	Males 12 600 Females 9600	From puberty onwards, people require 80-100 g of protein per day; 300 g of carbohydrate (except for very active people, who require more); 100 g of fat per day.
Adults Light work	Males 11 550 Females 9450	
Adults Moderate work	Males 12 100 Females 10 500	
Adults Heavy work	Males 17 500 Females 12 600	
Pregnant and nursing mothers	Pregnant 10 000 Nursing 11 300	Increased dietary calcium, iron and vitamins.

Figures from the Department of Health and Social Security (DHSS).

Children require less energy than adults because they have smaller bodies – fewer cells demanding energy. They eat more food than adults in relation to their size because of their higher basal metabolic rate and also because they are growing. Elderly people require fewer kilojoules as their basal metabolic rate is low and they are no longer growing.

Climate also affects a person's energy requirements. In cold climates more heat is lost to the external environment so more heat has to be generated within the body, although this is often compensated for by a less active lifestyle.

Milk. This is a secretion of the mammary glands, which are fully developed only in females. It is almost a complete food in itself as it contains all the essential substances of a human diet, with the exception of iron. It also contains many of the mother's own natural products such as antibodies (see page 182), which protect the newborn from infection.

Obesity. If the daily intake of food regularly exceeds the amount necessary to meet the body's energy needs, the body's stores of fat will gradually increase. An excess of body fat is known as obesity. This is not the same as being overweight. An individual who is overweight is simply heavier than the average person of the same height. The extra weight may be due to larger muscles or larger bones, as may be found in more athletic people. Obesity, on the other hand, is classified as a medical disease, and most doctors will offer patients treatment for it. The degree of obesity is assessed by measuring with a pair of calipers the thickness of a fold of skin just beneath the shoulder blades.

Obesity is regarded as a disease because it reduces a person's life expectancy. Excess weight increases the strain on bones and joints. Excess body fat also seems to increase the likelihood of contracting diabetes (see page 117), kidney disorder and heart disease. The simple remedy is to eat less.

People eat more than they should for a number of reasons. Firstly, with increasing age their rate of metabolism decreases, they become less active, so require less food. But many individuals fail to decrease their food intake as they get older, so develop 'middle-age spread'. Secondly, many people overeat as a result of a psychological disturbance or psychiatric disorder, and as a result get fat. Overeating is itself depressing, so they eat more to overcome the depression, so get fatter, and so it goes on. Thirdly, in some societies it is fashionable to overeat. Certain cultures look favourably upon people who are fat. But perhaps the most common cause of overeating is stress and the pressures of modern Western life. This tends to work against the regular taking of meals, which are readily replaced by lots of high-energy snacks. Drinking lots of beer and spirits can also lead to obesity, as alcohol is a high-energy food.

Food Tests

Tests for carbohydrates:
1. Naphthol test. Mix a few drops of 1 per cent alcoholic alpha-naphthol solution with a little of the test solution (thought to contain glucose, sucrose or starch). Then carefully pour about 5 mm^3 of concentrated sulphuric acid down the side of the tube. A violet coloration produced at the junction of the two liquids indicates the presence of a carbohydrate.

2. Test for reducing sugar (i.e. glucose, fructose and maltose). A little of the test solution is boiled with equal volumes of Fehling's A and B solutions. If a reducing sugar is present the solution will change from a clear blue colour to a bright orange colour.

3. Test for starch. To the food to be tested add three drops of a solution of iodine in potassium iodide. A blue-black coloration indicates the presence of starch. Heating the food makes the coloration disappear, but it reappears on cooling.

Tests for protein:
1. Millon's test. A little of the foodstuff is boiled with a few drops of Millon's reagent. Any particles of protein present will give a brick-red coloration.

2. Biuret test. Warm the food to be tested. Then add 2 mm^3 of 40 per cent sodium hydroxide solution and a few drops of 1 per cent copper sulphate solution. A violet coloration indicates the presence of protein.

Tests for fats:
1. Grease test. A simple test is that fats leave a translucent mark when rubbed or warmed on paper. This can be readily demonstrated with, for instance, butter, lard or castor oil.

2. Sudan III test. Place the food to be tested in a boiling tube containing water and boil. An oily layer will collect on the surface. Add a few drops of Sudan III reagent, shake and allow to settle. The oil will stain light red in the presence of fat.

3. Osmic acid test. To a small sample of the foodstuff add a few drops of osmic acid solution. With a positive result the sample will take on a brown-black appearance owing to the production of a fine precipitate of osmium oxide.

Digestion

Before food can be absorbed into the body, it must first be broken down into simple molecules, a process known as digestion. Digestion takes place in the gut by a series of physical and chemical processes, described below. Proteins are broken down into amino acids, fats into fatty acids and glycerol, and carbohydrates into simple sugars.

Chemical digestion is brought about by a range of digestive enzymes contained within digestive juices. Carbohydrates are broken down by enzymes called carbohydrases, fats by lipases and proteins are broken down by proteases.

The human gut or alimentary canal. Figure 12 showed the gut to be compartmentalized into a number of distinct sections. These and their associated organs are more accurately shown in Figure 13. The working of the gut will be explained by considering what happens to a typical meal as it passes from mouth to anus. The meal consists of the following foods;
- Grapefruit cocktail
- Lamb chops, potatoes, runner beans and gravy
- Sultana steam pudding with custard
- Cheese and biscuits
- Coffee with sugar and milk
- A bar of chocolate.

Carbohydrate	Protein	Fat	Others
Potatoes	Lamb chop	Lamb chop	Vitamin C – Grapefruit
Steam pudding	Cheese	Cheese	Sultanas
Chocolate	Milk	Milk	Roughage – Beans
Glacier cherry		Gravy	Water
Sugar			Various Minerals
Biscuits			
Custard			

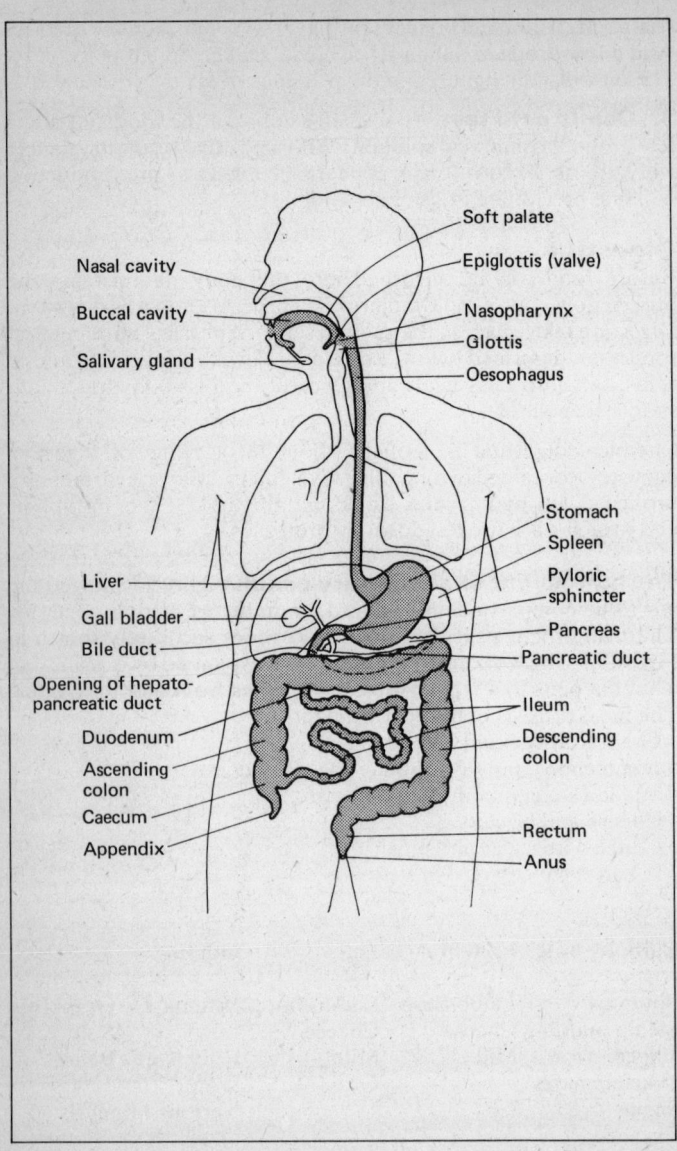

Figure 13. The human gut

The meal described on page 49 is fairly well balanced, although most of us should do without the bar of chococlate.

In the buccal cavity (mouth). Here the tougher foods such as the meat and beans are subjected to the physical cutting and crushing action of the teeth, the process of **mastication**. Teeth come in different shapes and sizes (see page 146). They are less important to humans than to other animals because human food is usually softened by cooking.

Chewing food, though, increases its surface area and mixes it with **saliva,** a digestive juice secreted by the salivary glands. The flow of saliva is started by the sight, smell or thought of food. It contains the carbohydrase, salivary amylase, which converts starch into maltose. (For a full list of gut enzymes and their action see Table 4). Saliva also lubricates the food in preparation for its movement into and down the oesophagus.

The tongue not only tastes the flavour of the food but is also used to shape it into a ball, the **bolus,** making it easy to swallow. Swallowing is a reflex action that is triggered by the food touching the soft palate and pharynx (see Figure 13). The opening to the larynx, the glottis, is covered by the epiglottis, a flap of tissue which prevents food from entering the lungs.

Into the stomach. After swallowing, the bolus moves down the oesophagus into the stomach. The oesophagus grips the food and propels it by a series of muscle contractions, the process of **peristalsis.**

Inside the stomach, the food is thoroughly mixed with gastric juice, which is secreted by the cells of the stomach lining. Gastric juice contains two enzymes, both proteases; **pepsin** which begins the digestion of proteins, and **rennin,** which acts on milk protein. Pepsin is a very destructive enzyme, so to protect the cells that secrete it, it is first produced in an inactive form, **pepsinogen.** This becomes activated by the very acidic environment of the stomach, created by the presence of hydrochloric acid.

The stomach itself is a tough muscular sac that on the entry of food begins to contract rhythmically. This pounds and churns the food into a semi-liquid state, **chyme.**

Gastric juice is produced following the sight, smell or taste of food but its flow increases greatly when food enters the stomach. This is brought about by the secretion directly into the bloodstream of a hormone, **gastrin,** from the stomach wall cells. Gastrin is carried along in the blood and stimulates the stomach to produce gastric juice.

Into the duodenum. By now, about an hour after the meal was eaten, the food is barely recognizable. It has been thoroughly broken up by the physical action of the teeth and stomach, but only the lamb chop, milk and potatoes have begun to be digested.

Chyme leaves the stomach through an opening bearing a sphincter muscle, the pyloric sphincter, and enters the duodenum. Here it is met by three separate digestive juices; **bile, pancreatic juice** and **intestinal juice.**

Bile is a viscous yellow-brown fluid, with the appearance of Castrol GTX motor oil. Each day about 1 litre of it is produced by the liver, from where it flows into the gall bladder (see page 39). Here it is stored and concentrated, much of its water being reabsorbed into the bloodstream. Bile flows from the gall bladder into the duodenum following the presence of acidified food from the stomach in this region of the gut. Bile contains a number of different substances, but chiefly bile salts, which reduce the surface tension of fats (see page 54), a process known as **emulsification.** This spreads out the fats to occupy a larger surface area so that enzymes in the pancreatic and intestinal juices can work on them. Bile salts also speed up the absorption of the fat-soluble vitamins A, D and K, and activate certain digestive enzymes. Bile itself contains no enzymes.

Pancreatic juice is, not surprisingly, produced by the pancreas, an organ that also secretes hormones (see page 116). It flows into the duodenum from the pancreatic duct. The main pancreatic enzymes are **trypsin,** which acts on proteins, pancreatic amylase and pancreatic lipase. Like pepsin, trypsin is produced in an inactive form so as not to damage the cells that produce it. This is called **trypsinogen,** and it is activated by the enzyme **enterokinase** contained in the intestinal juice. Pancreatic lipase breaks down fats (emulsified by the bile) into fatty acids and glycerol, which can be absorbed (see below).

Intestinal juice (succus entericus) is secreted by the inner wall of the duodenum and ileum. It is also rich in enzymes. These include maltase, which converts maltose into glucose; a number of proteases, which complete the digestion of protein; lactase and sucrase, which convert lactose and sucrose to glucose and galactose and glucose and fructose respectively; and enterokinase, which activates trypsin.

Absorption The products of digestion are now in a form small enough to be taken into the body. Absorption takes place in the ileum, which is very long and, by virtue of its highly folded inner surface (see Figure 14), has an extremely large surface area.

On the inner surface of the ileum are numerous finger-like projections known as villi (singular **villus**). Within each villus is a dense network of blood capillaries that originate from the mesenteric artery and drain into the hepatic portal vein. In the centre of each villus is a **lacteal,** which contains lymph.

The digested meal is now a solution of amino acids, monosaccharides, fatty acids and glycerol, vitamins and mineral salts.

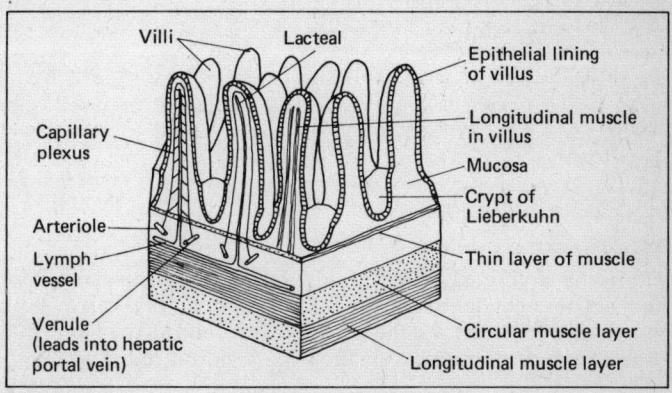

Figure 14. Section through the wall of the ileum

The amino acids and sugars are absorbed by a combination of diffusion and active transport across the epithelium of the villi into the capillaries beneath. The fatty acids and glycerol are absorbed into the epithelial cells where they are re-synthesized into fat globules and shed into the lacteals. From the lacteals the globules enter the main lymphatic system, eventually passing into the bloodstream where lymph drains into the veins near the heart (see page 90).

Mineral salts, vitamins and some water are also absorbed in the ileum. The undigested remains of the meal pass on to the large intestine where, in the colon, much of its water is removed. By the time it reaches the rectum, the undigested food is semi-solid again and is in a form ready to be eliminated from the body through the anus; this constitutes the **faeces.**

Table 4. Human digestive juices

Secretion	Source	Site of action	Enzymes	Substrate	Products
Saliva (neutral)	Salivary glands	Mouth	Amylase	Starch	Maltose
Gastric juice (very acid)	Stomach wall	Stomach	Hydrochloric acid (not an enzyme)	Pepsinogen	Pepsin
			pepsin, rennin	Proteins	Polypeptides
Bile (alkaline)	Liver	Duodenum	Bile salts (not enzymes)	Fats	Fat droplets
Pancreatic juice (alkaline)	Pancreas	Duodenum	Amylase	Starch	Maltose
			Trypsin	Protein	Polypeptides
			Peptidases	Polypeptides	Amino acids
			Lipase	Fats	Fatty acids and glycerol
Intestinal juice (alkaline)	Wall of duodenum and ileum	Duodenum and ileum	Enterokinase	Trypsinogen	Trypsin
			Amylase	Starch	Maltose
			Maltase	Maltose	Glucose
			Sucrase	Sucrose	Glucose+fructose
			Lactase	Lactose	Glucose+galactose
			Peptidases	Polypeptides	Amino acids

The liver

This is the largest organ in the body. It is reddish-brown in colour and lies immediately beneath the diaphragm. It receives both blood from the heart, via the hepatic artery, and all the blood that drains from the intestines, via the hepatic portal vein.

The liver has more functions than any other organ. The most major of these is to control which substances enter the main blood circulation – and therefore reach all the cells of the body – from the gut. Blood in the hepatic portal vein contains everything that is absorbed by the body from the gut, with the exception of one or two substances such as alcohol, which are absorbed into the bloodstream in the stomach. As such, the liver is in a key position. It is able to act in a manner similar to that of customs officials at a country's border. These officials check visitors' identities and their luggage and all the freight that enters the country. The liver selects out from all the absorbed substances which ones will be let into the circulation and in what quantities. The structure of the liver has evolved such that it carries out this selection task particularly efficiently (see Figure 15).

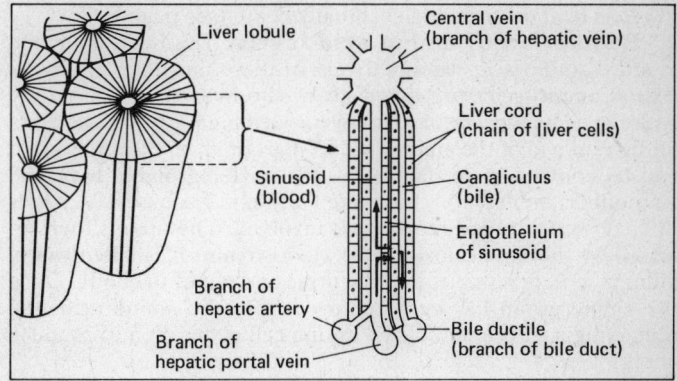

Figure 15. Structure of a liver lobule

The liver is composed of thousands of cylindrical liver lobules each about 1 mm in diameter. The lobules are packed with liver cells that radiate out in rows from the centre to the periphery. Running alongside each strip of liver cells are branches of the hepatic artery, hepatic portal vein and the bile duct. In the centre of each lobule is a branch of the hepatic vein, which is connected to the other vessels by a fine channel, a sinusoid.

Oxygenated blood from the heart enters a sinusoid from the hepatic artery and it immediately becomes mixed with blood mixed with food, this entering from the hepatic portal vein. The blood then flows down the sinusoid and into the hepatic vein. As it does so, the liver cells take up the substances they require and shed into the blood their products. The exception to this is the bile, which is shed into the adjacent bile channel (the canaliculus). All of the liver's functions are performed on the blood by these liver cells as it flows by.

These functions are;

1. Regulation of blood sugar levels. Liver cells take up excess glucose from the blood and convert it to the polysaccharide glycogen, which is insoluble. Glycogen is stored in the liver cells until the level of glucose in the blood begins to drop. It is then converted back into glucose and released into the blood as it passes by. Uptake of glucose by liver cells is stimulated by the hormone insulin, which is secreted steadily by the pancreas every day. Without insulin the level of blood glucose will vary wildly. This can lead to coma and eventually death (see page 116).

2. Regulation of amino acid levels. The body is unable to store amino acids so any excess of these must be destroyed. Excess amino acids are taken up by the liver cells where they undergo a chemical process known as deamination. This consists of the removal of the amine ($-NH_2$) group from the amino acid and its conversion to ammonia (NH_3). Being highly toxic the ammonia is immediately converted into non-toxic urea. A series of enzyme controlled reactions is involved. The urea is then released by the liver cells into the bloodstream. It is taken to the kidneys, where it passes into the urine and out of the body. Once the amine group has been removed from the amino acid, the remaining acid is metabolized during cell respiration to produce energy.

3. Detoxification. The liver is able to remove from the blood any toxic substances that may have been absorbed from the gut. It can convert water-insoluble substances to water-soluble substances which can be excreted into the bile or urine and eliminated. Drugs and hormones are also inactivated by the liver.

4. Elimination of erythrocytes. Exhausted erythrocytes (red blood cells) are taken up by phagocytic cells lining the liver sinusoids. The haemoglobin is broken down by liver cells into a green pigment, biliverdin, and a brown pigment, bilirubin. These are removed into the bile duct and are responsible for the appearance of the bile.

5. Production of bile. Bile is a type of waste product of liver cell activity and consists of a variety of pigments and salts. As we have seen, bile salts play a part in fat digestion.

6. Storing blood. The veins in the liver have a great power to expand and contract. This enables the liver to act as a store of blood, releasing it into the circulation at times of urgent need, such as during exercise.

7. Storage of vitamins. The fat-soluble vitamins, A and D, are stored inside liver cells. This explains why animal liver is a good source of these vitamins for a human diet.

8. Production of heat. During respiration the liver cells generate much heat energy which warms the blood.

The role of the liver in homeostasis

The liver is essentially a regulatory organ, which means that it attempts to regulate, or control, the level of certain substances in the blood within strict limits. The control of blood sugars for example, is crucial for the correct functioning of the body's cells. They need a constant supply of sugar for respiration, 24 hours a day. Without the action of the liver the blood would only contain sugar following a meal. This would not do.

You should realize by now that sugar is not the only substance or **parameter,** that needs to be strictly controlled by the body. Even very small fluctuations in body temperature, osmotic pressure, pH or amounts of other chemical substances will disrupt cell metabolism, and in many cases will kill off cells altogether.

This was clearly apparent to the French physiologist, Claude Bernard, back in 1857, when he wrote that 'the constancy of the internal environment is the condition for free life'. What he meant was that if certain aspects (parameters) of the internal environment (see page 34), were not controlled within very strict limits, the body's cells would die. Bernard coined the term **homeostasis** to refer to the processes that are used to ensure a constant internal environment. It literally means 'the same state'.

The most important aspects of the internal environment that must be kept constant are:

1. Its chemical constituents, i.e. glucose, mineral ions, etc.
2. Its temperature.
3. Its osmotic pressure.
4. Its level of carbon dioxide.

The principles of homeostasis can be illustrated by looking generally at how the liver is involved in the maintenance of the level of blood sugar. Maintenance of body temperature is dealt with on page 99, and that of osmotic pressure in Chapter 5. Maintenance of CO_2 levels is dealt with in Chapter 3.

The level of blood sugar varies from its normal value (norm) after a meal, when it tends to rise, and following physical activity, when it tends to fall. Figure 17 shows how corrective mechanisms operate to return the blood sugar to the norm. (Full details of the action of insulin are given on page 116). This illustrates an important aspect of homeostasis, **negative feedback.** In other words, inform-

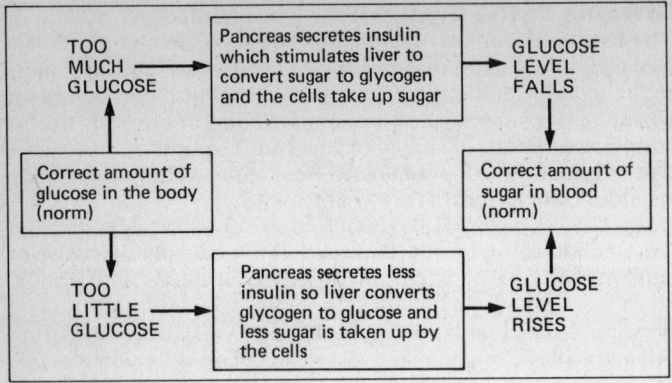

Figure 16. Homeostatic control of blood sugar

ation about the level of blood sugar is constantly fed back to the system that initiates corrective mechanisms so that any deviation from the norm is avoided. This is illustrated below. This basic principle of homeostatic control applies to all the other parameters that need to be controlled by the body

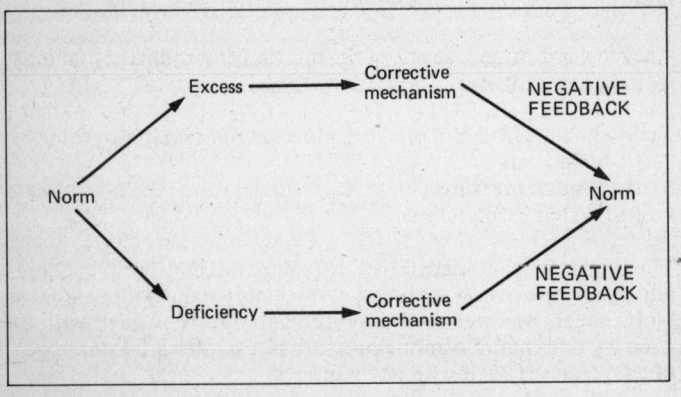

Figure 17. Negative feedback

Key terms
Absorption Uptake of soluble substances from the gut into the blood.
Anion Negatively charged particle.
Bolus Round ball of food prepared by the tongue for swallowing.

Cation Positively charged particle.
Chyme Liquid product of digestion by the stomach.
Deamination Breakdown of excess amino acids by removal of amine group.
Defaecation Removal of undigested food from the gut; also known as egestion.
Detoxification Removal of harmful substances or organisms from the blood by the liver.
Diet The types and quantities of food eaten by a person.
Digestion Process of breaking down complex insoluble food into simple soluble substances which can be absorbed into the body.
Energy The capacity to do work.
Enzymes Proteins that speed up the rate of chemical reactions.
Fibre Undigestable food which forms the bulk of a meal.
Gut Long compartmentalized tube that processes food.
Homeostasis The process of maintaining a constant internal environment.
Mastication Crushing action of teeth that breaks up food.
Metabolic rate The rate at which the body's chemical reactions take place.
Negative feedback Self-regulating process which controls body parameters.
Obesity Excess of body fat.
Parameter A measurable aspect of the body's activity, e.g. temperature, chemical levels in the blood.
Peristalsis Muscular movement of food along the gut.
Scurvy A deficiency disease caused by lack of vitamin C in the diet.
Vitamin Essential ingredient in the diet, lack of which causes a specific disease.

Chapter 3
The Lungs and Ventilation

Ventilation is the process by which a living organism obtains oxygen from its external environment. Man requires oxygen for cell respiration.

The terms **ventilation** and **respiration** are often confused. Respiration is now more usually and correctly used to refer to the chemical processes that release energy from food within a cell, i.e. cell respiration (see page 29). Ventilation is sometimes called external respiration to distinguish it from internal or cell respiration.

In a simple unicellular organism such as *Amoeba*, ventilation is a very straightforward process. Oxygen dissolved in the surrounding water is at a greater concentration than that dissolved in the cell cytoplasm. Thus oxygen diffuses from the water into the organism (see page 27).

However, oxygen cannot diffuse through the many layers of cells of a multicellular organism such as man. So in common with many other higher animals man has evolved special organs, the lungs, by which to ventilate the body. Ventilation is achieved as air is forced in and out of the lungs by means of repeated contraction and relaxation of the diaphragm and the muscles between the ribs, the intercostal muscles (Figure 18). This process is known as breathing.

Structure of the human ventilation system

The lungs are located within the thorax and are protected by the ribcage. At the base of the thorax is a tough sheet of muscle, the **diaphragm,** which separates the thorax from the abdomen. The space between the lungs and the ribcage is fluid-filled and comprises the **pleural cavity.** This is lined by the pleural membranes, which keep the cavity air-tight and secrete a fluid that lubricates the surface of the lungs to facilitate their continuous movement. Pleurisy is an inflammation of the pleural membranes due to infection by bacteria or viruses.

The lungs are connected to the outside by the **trachea,** or windpipe, nose and mouth. The trachea is a flexible tube that is reinforced with rings of cartilage. It leads from the pharynx to a point just above the heart, where it divides into two branches, the bronchii, one branch going to each lung.

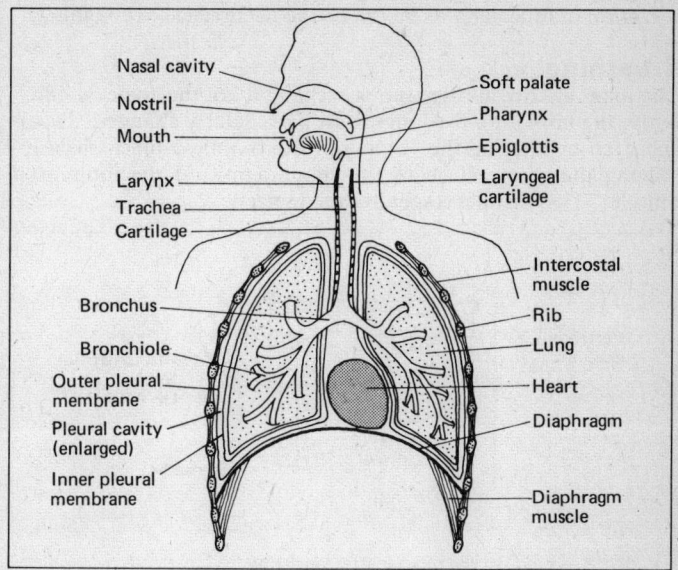

Figure 18. Section through the head and thorax

The **bronchii** (singular bronchus) themselves each divide to form a multitude of tiny tubes known as **bronchioles.** These eventually end at multi-lobed thin-walled sacs, the alveoli. The lungs are therefore very elaborate networks of tubes with ultimately only one way in or out.

Both the trachea and the oesophagus lead from the pharynx, so a mechanism is required to prevent food from entering the lungs. This is the function of the **epiglottis,** a small flap of cartilage which covers the glottis during swallowing.

Air contains particles of matter, or dust, which could collect in the lungs and interfere with their action. To stop this happening, the nasal cavity, trachea and bronchii are lined with glandular epithelial tissue. The cells of the lining secrete mucus, which traps the dust, and the rhythmic action of their cilia causes the trapped particles to pass back up the respiratory tract. Alternatively a coughing reflex ejects the mucus from the glottis and this may be spat out or swallowed. Inflammation of the respiratory system lining by chemicals or micro-organisms often results in the copious secretion of mucous; this increases the occurrence of coughing.

Breathing

The lungs are organs of gaseous exchange. For this to occur efficiently, the air within the lungs must be regularly changed. This is achieved by altering the volume of the thorax, a mechanism involving the joint action of the diaphragm and the intercostal muscles. Two distinct stages are recognized.

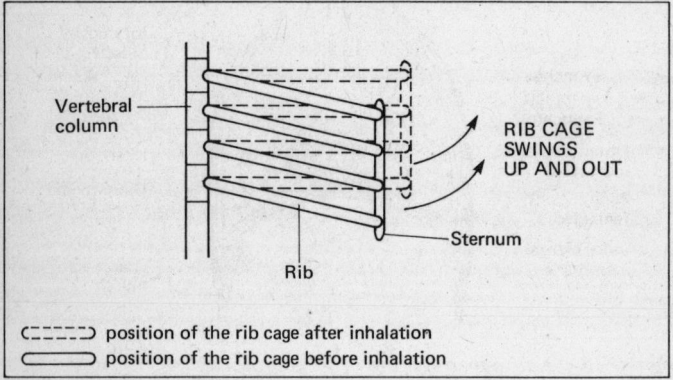

Figure 19. Movement of the ribcage at inhalation

1. Inhalation. For air to pass into the lungs the volume of the thorax must be increased so that the suction pull on the lungs is greater than normal. As the lungs are made to expand, air pressure within the lungs becomes lower than that outside the body. As different levels always tend to equalize themselves air will rush into the lungs from outside.

The thoracic volume is increased firstly by a contraction of the diaphragm, which becomes flattened from its normal domed position (see Figure 18). Secondly, and at the same time, a contraction of the intercostal muscles causes the ribcage to swing up and out.

2. Exhalation. Exhalation is more of a passive process. The intercostal muscles relax causing the ribcage to return to its original position under the force of gravity. Similarly, the diaphragm relaxes and resumes its domed position. This is assisted by a gentle pressure exerted on the diaphragm by the abdominal organs below. The net result is a decrease in thoracic volume and a recoil of the elastic lung tissue with a subsequent increase in air pressure within the lungs. So air is forced out of the lungs until pressures inside and outside the organs are once again equal.

The lungs will not expand during inhalation unless the surrounding chest wall is air-tight. This depends upon the integrity of the outer pleural membranes. Should these be punctured, for example, by a bullet or a broken rib, the lungs will collapse and no longer function until the wound is healed.

In an average person the amount of air inhaled and exhaled during one breath is about 500 cm^3. This figure is known as the **tidal volume,** or tidal air, and it can be measured by breathing in and out of a device known as a **spirometer** (Figure 20).

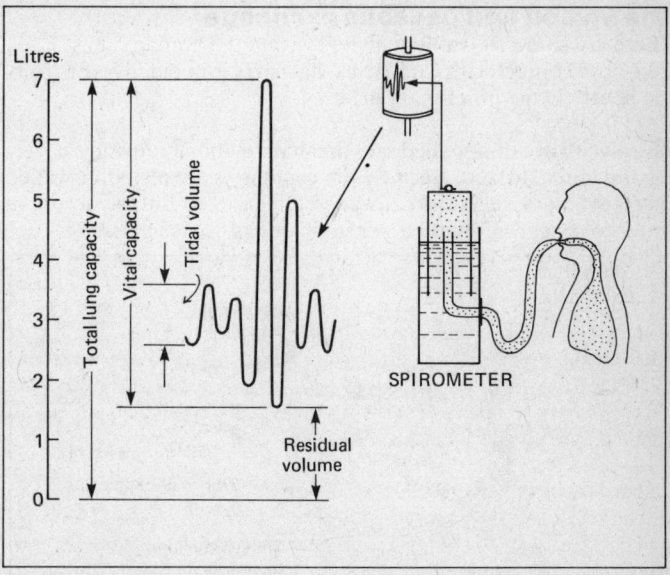

Figure 20. Diagram of spirometer in use on an athlete

At rest, the tidal volume remains fairly constant. But if you take a deep breath, a litre of air over and above the tidal volume can be breathed in. This is known as the **inspiratory reserve volume.** Similarly if you try to expel as much air as possible, an extra litre of air can be breathed out. This is known as the **expiratory reserve volume.**

There is a further volume of air that fills the trachea and bronchii which never reaches the alveoli before it is expelled during exhalation. This volume is known as the **dead space** and in an average person is about 170 cm^3.

The tidal volume, the two reserve volumes and the dead space are together known as the **vital capacity.** In an average person the vital capacity is almost 4 litres, but in a trained athlete it may be as much as 5½ litres.

Even after maximum exhalation, a volume of air remains in the lungs, this amounting to about 1600 cm^3. This volume is known as the residual volume. The residual volume plus the vital capacity comprise the **total lung capacity,** i.e. about 5.6 litres.

The alveoli and gaseous exchange

There are some 700 million alveoli within the human lungs. Each is in close contact with capillaries that carry blood delivered from the heart via the pulmonary arteries.

The alveoli are thin-walled sacs that have a film of moisture covering the inner surface. Because the capillaries are also thin-walled there is only a very short distance for gases to diffuse across in order to pass from blood to air in the lungs or vice versa.

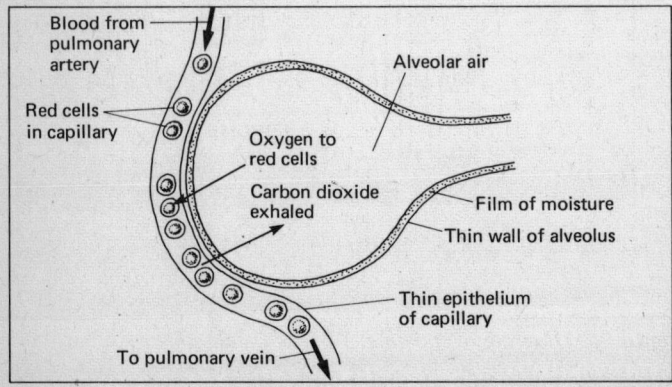

Figure 21. Exchange of gases in alveolus

After inhalation, the oxygen concentration in air in the space within each alveolus is high compared to that in the blood. So oxygen dissolves in the alveolar film of moisture and diffuses down a concentration gradient into the bloodstream. Once in the blood the oxygen diffuses into the red blood cells. Here it combines with haemoglobin to form oxyhaemoglobin. This removes it from solution in blood and maintains the concentration gradient for further diffusion of oxygen.

Similarly there is a carbon dioxide concentration gradient, but in the opposite direction. The inhaled air is low in carbon dioxide relative to the blood. So this gas diffuses out of the blood, crosses the capillary and alveolar walls, and is removed during exhalation (see also Chapter 4).

The blood thus oxygenated is carried away from the lungs via the pulmonary vein. This conducts the blood to the left atrium of the heart. From here it passes into the left ventricle, which then pumps it to the rest of the body.

The union of oxygen and haemoglobin in the red blood cells is a very loose one. Oxygen molecules are quickly taken up in the lungs and equally readily detached in the tissues. Each haemoglobin molecule carries four oxygen molecules, so we may write:

$$Hb + 4O_2 \underset{\text{Tissues}}{\overset{\text{Lungs}}{\rightleftarrows}} HbO_8$$

Haemoglobin   Oxyhaemoglobin

When the red cells pass tissues which have low oxygen concentrations, oxygen is released from the oxyhaemoglobin. The gas then diffuses into the tissues and is used in cell respiration. At the same time, carbon dioxide produced as a waste product of cell respiration is taken up by the blood for transport back to the lungs.

Control of oxygen and carbon dioxide levels in blood

The structure and organization of the human ventilation (respiratory) system together ensure a high capacity for gaseous exchange. But the system must be able to respond to changes in the body's demand for oxygen.

This is quite simply achieved by altering the breathing rate and heart rate as necessary. The breathing rate is controlled by a group of cells in the medulla oblongata of the brain (see page 111). They make up the **respiratory centre,** and from them nerves convey impulses to the diaphragm and intercostal muscles. The heart rate is controlled by the **cardiovascular centre,** also found in the medulla oblongata. Nerves from here carry impulses to the heart's **pacemaker** (see page 86). The two centres respond appropriately to the level of carbon dioxide and, to a lesser extent, oxygen, in the bloodstream. If the concentration of carbon dioxide rises, the centres respond by increasing the breathing rate and heart rate. If the carbon dioxide level falls, the centres cause a decrease in the two rates.

The nerve cells in both centres are constantly being informed of the carbon dioxide level in the blood by sensory cells (**chemoreceptors** sensitive to carbon dioxide) at the base of the internal carotid arteries of the neck. These chemoreceptors form the **carotid body.** If the carbon dioxide level in the blood rises, the cells send more nerve impulses than normal to the respiratory and cardiovascular centres in the brain. The centres then respond as described.

It should also be noted that via the higher brain centres one is able to influence breathing rate consciously. Hence you can hold your breath or breathe deeply as a matter of choice. But once the decision to do so is forgotten, this override system ceases and the medulla reverts to normal functioning.

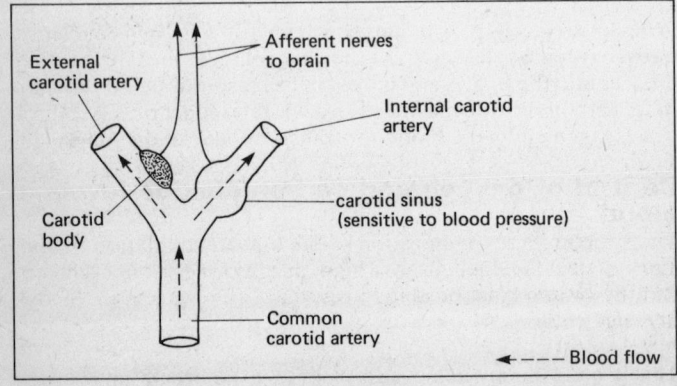

Figure 22. The carotid body and sinus

There is a steady stream of impulses from the respiratory centre enabling one to sleep without having to worry about breathing. Occasionally, however, this centre may be damaged or rendered ineffective. For example, there is a condition where the respiratory centre is defective at birth. The victim must breathe by a conscious effort of will and must sleep in an **iron lung** (see Figure 23). This machine, which automatically moves the ribcage of the patient, is also used where the respiratory centre has been damaged, for example, in a road accident.

Alternatively, the breathing mechanism may not function because of damage to the intercostal muscles, the diaphragm or to the nerves leading to them. The latter happens in some cases of poliomyelitis. Again, an iron lung may be the only way to save the victim.

Artificial respiration

An iron lung is an example of an artificial respirator, that is a device that will force air into the lungs when the natural breathing mechanisms break down.

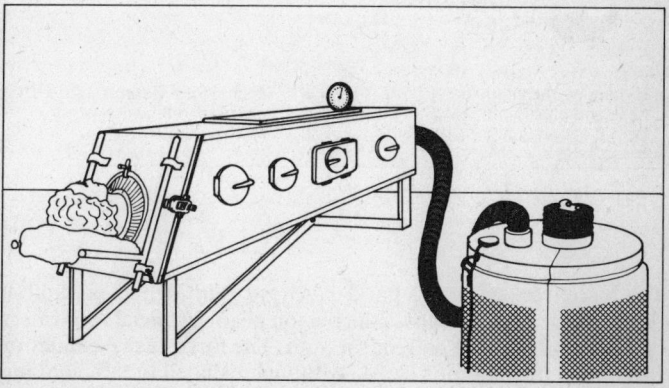

Figure 23. An iron lung in operation

It consists of a large air-tight cabinet in which the patient lies, his head sticking out through a flexible collar. The inside of the cabinet is connected to an electronically motor-driven bellows. At each stroke of the bellows a fixed amount of air is withdrawn from the cabinet. The resulting fall in pressure produces a suction pull on the patient's chest, which thus expands. This in turn produces suction on the lungs so that air is drawn in. As the bellows collapse, the procedure is reversed. The action of the bellows is therefore directly comparable to the action of the diaphragm and intercostal muscles during normal breathing.

Artificial respiration can also be applied to people without the aid of a gadget such as an iron lung. An example is the mouth-to-mouth technique of respiration that is mentioned in the Bible (II Kings, iv, 34) and which is still used today to resuscitate patients who have stopped breathing through drowning, gas poisoning or electrocution.

Extending the victim's head and pushing forward his jaw before applying mouth to mouth breathing

Pinching the victim's nostrils and blowing gently into his mouth

Figure 24. Artificial respiration

The patient is laid on his back. His head is tilted back and, after pinching the patient's nose, the person giving artificial respiration breathes out into the patient's mouth. The force of the exhalation will inflate the patient's chest. Although exhaled breath contains less oxygen than inhaled atmospheric air, it provides enough oxygen for the patient's needs until his own lungs can function again.

Inhaled and exhaled air

The net result of ventilation and gaseous exchange is that inhaled air is different in composition from exhaled air. These differences can be seen opposite:

Approximate composition of inhaled and exhaled air

Constituent	Inhaled air (atmospheric air)	Exhaled air
Oxygen	21%	16%
Carbon dioxide	0.03%	4%
Nitrogen	79%	79%
Water vapour	Varies	Always saturated
Other differences:	Often cool; may contain dust	Warm; most dust removed

Breathing through the nose increases the moisture content and temperature, and decreases the dust content, of air reaching the lungs. (In the nasal passages, serous glands secrete a fluid that moistens the air; mucous glands produce mucus that traps dust particles; and the warmth of blood in capillaries raises the temperature of the air.) This is important on very cold and/or dry days. Too much dry air could effect the moisture surface of the alveoli of the lungs to the point where they could not carry out effective gaseous exchange. And too much cold air could reduce the metabolic rate of lung cells and, in extreme circumstances, could overcool the whole thorax. So whenever possible nose breathing is always to be preferred to mouth breathing.

The effects of exercise

During exercise a large number of changes occur within the body. There is a general constriction of arterioles (except to vital organs), the spleen (the body's main blood reservoir) contracts, metabolic rate increases, body temperature rises, and the heart and breathing rates increase. These are all brought about by action of the hormone **adrenaline** (see page 119).

The last two changes ensure that gaseous exchange occurs sufficiently quickly to meet the body's demand for extra oxygen. But this may be jeopardized if the lungs or circulatory system are overworked (or damaged). The person exercising will be able to increase his effort only until the capacity of the ventilation mechanism is reached. If he continues to increase his effort the body tissues will not receive enough oxygen and will then perform anaerobic respiration (see page 30), resulting in the production of lactic acid.

After he has stopped activity, the lactic acid is transported to the liver where it is converted back to glycogen. But this chemical process requires oxygen. It constitutes the so-called **oxygen debt,** and accounts for the heavy panting that follows intense exercise, an attempt to quickly take in large quantities of oxygen.

The effect of pollutants on the lungs

How soon the development of an oxygen debt during exercise begins depends also upon the health of the subject's lungs. The efficiency of the lungs can be impaired by a wide variety of pollutants, the most common of which is tobacco smoke.

Like all smoke, that from a cigarette, cigar or pipe consists of fine particulate matter suspended in a hot gas. Once taken into the lungs, much of the gas condenses and remains there, as do many of the particles. That a change in composition of the smoke occurs can be seen by comparing the colour of smoke from the end of a cigarette with that exhaled by a smoker. The difference represents the residues left in the smoker's lungs.

The tar and other chemicals that are deposited in the lungs by smoking line the alveoli and impair gaseous exchange. They can also cause severe lung inflammation, which eventually causes emphysema – a progressive degeneration of lung tissue. In addition, inflammation of the bronchi can occur **(bronchitis).**

For some unfortunate smokers, the constant irritation to the lung lining results in serious cellular changes, which can lead to lung cancer.

Smoking may also cause changes in the circulatory system, leading to heart disease and high blood pressure.

Other pollutants of the lungs include chemicals in the exhaust smoke of car engines and heavy dust concentrations in the atmosphere. The latter can be a particular problem in some factories and regulations have now been brought in to control these. In addition, blue asbestos, the dust particles of which cause highly malignant lung cancer, has now been banned in Great Britain.

Dust also causes coughing, which is a spasmodic reflex contraction of the intercostal muscles and diaphragm. Its effect is to cause an almost explosive exhalation that is designed to sweep the irritating material from the lungs. If the offending material is not removed the first time, the coughing continues. With diseases such as asbestosis, coughing bouts can last many minutes, This can become quite serious should the victim become totally exhausted through the strain of coughing; he may even die as a result.

If the lining of the nasal passages, the nasal mucosa, is irritated by dust, a similar explosive reflex exhalation occurs, sneezing.

Coughing, sneezing, and a copious production of respiratory mucus, can be triggered by the pollen of plants. This is because some humans are **allergic** to the proteins and waxes that make up the outer coats of pollen grains; their bodies treat these chemicals as 'foreign' and produce antibodies against them (see page 182). The result is a well-recognized condition, hay-fever.

Key terms
Allergy Abnormal immune reaction to a foreign substance.
Artificial respiration Forcing air into the lungs when natural breathing stops.
Breathing Movements of the body to force air in and out of the lungs.
Bronchitis Inflammation of the bronchi.
Diffusion Movement of molecules in a gas or liquid from a region of high concentration to a region of low concentration across a semi-permeable membrane.
Exhalation Breathing out.
Inhalation Breathing in.
Respiration Strictly process in which energy is produced in cells.
    Aerobic: oxygen is used to break down sugar completely (to carbon dioxide and water).
    Anaerobic: sugar is only partly broken down (to lactic acid) in absence of oxygen.
Respiratory surface Where animals exchange respiratory gases with the environment; in man, the walls of the alveoli.
Resuscitation Revival of the vital processes of the body.
Spirometer Instrument that measures the parameters of ventilation.
Tidal volume Air moved in and out of lungs during breathing.
Ventilation Process of obtaining air from the external environment.

Chapter 4

Blood: the Body's Transport System

The function of blood is to carry substances. It provides the body with an internal transport system, rather as London's buses and tubes provide Londoners with a means of travelling about. London, like the body, would grind to a halt without its transport system.

Blood is contained by the body's vascular system, which consists of a continuous series of tubes, or vessels, of varying thickness and diameter. The larger tubes are the arteries and veins, the smaller ones, the venules, arterioles and capillaries.

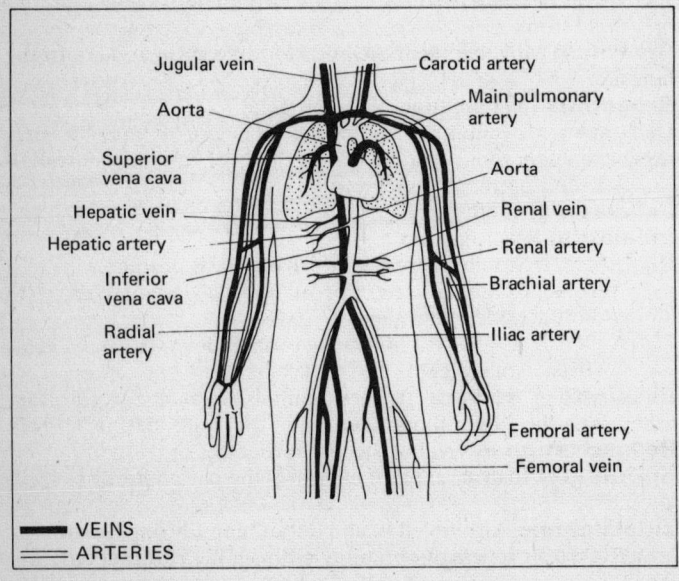

Figure 25. The heart, the main arteries and veins

Blood is pumped through the tubes by a muscular pump, the heart. The flow of blood is continuous and is in one direction only.

Structure of blood

Blood consists of a fluid known as **plasma** in which are suspended three types of cell; red blood cells **(erythrocytes)**, white blood cells **(leucocytes)** and cell fragments **(platelets)**.

1. Plasma. This is the straw-coloured aqueous component of blood. It is made up largely of water in which is dissolved a number of important substances; glucose, amino acids, sodium chloride, hormones, bicarbonates, urea and a number of plasma proteins, some of which aid in the clotting of blood (see page 77). One of these clotting proteins is called fibrinogen. Plasma with its fibrinogen removed is know as **serum.**

2. Erythrocytes. The red blood cells are small biconcave discs. There are approximately 5 million per mm^3 of blood. Each cell is surrounded by a thin plasma membrane and packed full of the iron-containing protein **haemoglobin**. There is no real need for a nucleus, which is lost early on in the cell's development in the bone marrow. But without a nucleus an erythrocyte has a short, finite life span, usually about 120 days. After this time the cells are broken down by the liver and spleen. Their lack of a nucleus also gives rise to their biconcave appearance, the plasma membrane being depressed in the region where there would normally be a nucleus.

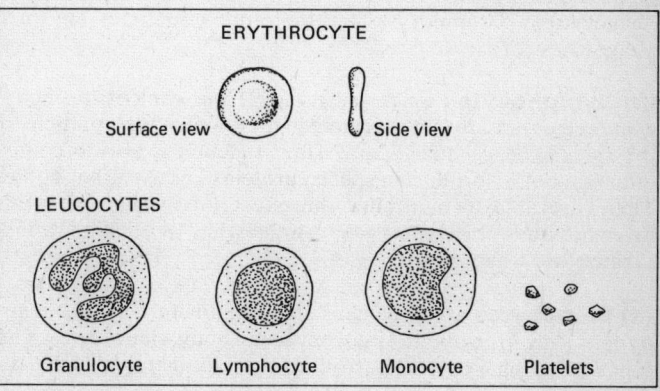

Figure 26. Cellular components of blood

The function of a red blood cell is to carry oxygen round the body. Oxygen is collected in the lungs, combining with the haemoglobin to form **oxyhaemoglobin.** In regions where oxygen is required, oxyhaemoglobin releases the gas to the tissues (see page 64).

Haemoglobin is also able to combine with carbon monoxide, a poisonous gas present in cigarette smoke and car exhaust fumes. The **carboxyhaemoglobin** that is produced effectively reduces the amount of haemoglobin that is available to combine with oxygen. The body therefore suffers a temporary or permanent state of **anoxia,** which is a shortage of oxygen in the body. This explains why during exercise, when extra oxygen is required, some smokers are often short of breath and why a garage or car full of exhaust fumes can prove fatal.

3. Leucocytes. The white blood cells are larger, colourless cells. And unlike red cells they vary enormously in size and shape according to the function they perform. They also have a nucleus. There is approximately only one leucocyte to every 600 red cells, i.e. 8,000 per mm^3 of blood. There are three main types of leucocytes:

(a) Granulocytes are irregular-shaped cells formed in the bone marrow. They make up about 75 per cent of the body's leucocytes. Their function is to protect the body from infection by ingesting foreign material or micro-organisms in the blood and tissue fluid (see Chapter 9). This process of ingestion by white cells is called phagocytosis. Granulocytes are therefore sometimes known as phagocytes.

(b) Lymphocytes make up about 21 per cent of the body's white cells. They are manufactured in the lymph nodes, particularly the spleen, hence their name. Their function is also to combat infection but by producing special proteins known as antibodies. These are able to neutralize the effects of foreign chemicals, foreign tissues or invading micro-organsisms by sticking to their surface (see Chapter 9).

(c) Monocytes are large cells also manufactured in the lymph nodes. They form about 4 per cent of the body's leucocytes. They function as phagocytes, patrolling the bloodstream and tissue fluids.

4. Platelets. When blood is exposed to the air as the result of a cut or wound, a complex chemical sequence causes it to clot. This useful process prevents excess loss of blood (known as a **haem-**

orrhage) and prevents micro-organisms from entering the wound. For this to occur cell fragments called platelets must be present in the blood. Platelets are produced in the bone marrow and do not have an intact plasma membrane or a nucleus. There are about ½ million of them per mm^3 of blood. Their sole function is to bring about the blood-clotting mechanism (see page 77).

Functions of blood
1. Replenishing tissue fluid. The primary function of blood is to provide tissue fluid (lymph) and hence all cells with their requirements. As we saw at the end of Chapter 1, each cell in the body is bathed in this watery tissue fluid. Chemical supplies are filtered out of the bloodstream into the fluid, from where they are absorbed by the cells. Waste products are transferred from the cells via the tissue fluid to the blood.

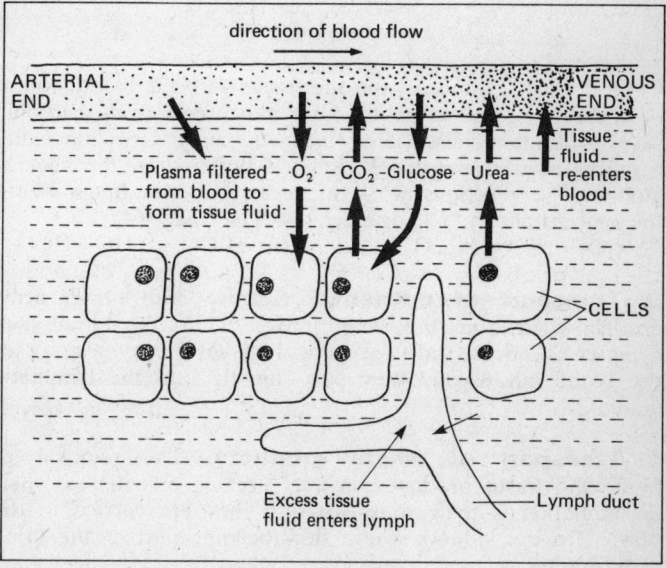

Figure 27. Blood, tissue fluid and lymph

If at any time the tissue fluid contains inadequate supplies of oxygen or nutrients, contains poisonous chemical substances, is at too high or low a temperature or osmotic pressure, the surrounding cells will be unable to function and will die.

2. Transport of oxygen and carbon dioxide. The biconcave shape of an erythrocyte provides a large surface area for the uptake of oxygen in the lungs. Its lack of a nucleus permits more haemoglobin to be packed into the cell than would normally be possible.

The haemoglobin molecule has four iron-containing haem groups, each of which can combine with an oxygen molecule. Oxygen uptake occurs readily across the alveoli in the lungs (see Chapter 3), where the oxygen concentration is high. Conversely, in the tissues, where oxygen concentration is low, the unstable oxyhaemoglobin dissociates releasing the oxygen to the tissues.

$$Hb + 4O_2 \underset{\text{Tissues}}{\overset{\text{Lungs}}{\rightleftharpoons}} HbO_8$$

Carbon dioxide diffuses from the tissues into the blood, where it combines with water to form the weak acid, carbonic acid. This then dissociates into its constituent ions; the resulting acidity (due to H^+ ions) is buffered by the haemoglobin. The reactions work in reverse in the lungs where the concentration of CO_2 is much lower (see page 64).
(i) $CO_2 + H_2O = H_2CO_3$ (ii) $H_2CO_3 \rightleftharpoons H^+ + HCO_3^-$

3. Transport of nutrients. Glucose and amino acids are absorbed from the small intestine into the blood (see Chapter 2). Some fatty acids and glycerol may also enter the blood but usually they pass directly into the lymphatic system.

4. Transport of waste products. The removal of metabolic waste products such as urea is essential as their accumulation is toxic to the body. They are carried in the plasma to the kidneys where they become part of the urine (see Chapter 5).

5. Transport of hormones. Hormones are secreted by endocrine glands directly into the bloodstream for transport to their target organ(s) (see Chapter 6).

6. Heat regulation. Blood plays an important part in regulating the body's temperature. It distributes evenly heat produced by the metabolic activity of cells. Also certain blood vessels are able to dilate and constrict, thereby increasing or decreasing the flow of blood to the skin. Heat can dissipate readily from the skin so when extra heat has to be lost from the body, more blood is pumped to the skin. When heat needs to be conserved, blood flow to the skin is restricted (see page 99).

7. Defence against disease. Blood has an important role in the body's highly complex systems of defence that protect it from infection (see Chapter 9).

The body's first line of defence against harmful micro-organisms is the passive mechanical protection provided by the skin. Where the skin is naturally broken, i.e. at the eyes, mouth and anus, nose, ears and genitals, additional protective secretions (such as tears) are provided. However, if the skin is cut or its protective secretions disturbed in any way, then micro-organisms can easily enter the body and infection results.

Blood clotting. The first barrier to wound infection is provided by the ability of the blood to clot. This plugs the wound. The mechanism involved is as follows (see Figure 28):

1. In the presence of air, platelets release the chemical **thrombokinase** (also called thromboplastin).
2. In the presence of calcium ions, thrombokinase converts the blood protein **prothrombin** into the enzyme **thrombin.**
3. Thrombin converts the soluble plasma protein **fibrinogen** into an insoluble network of threads called **fibrin.** This network traps the blood cells causing them to clot.

Certain individuals have blood which will not clot when exposed to the air. Somewhere in the series of chemical reactions described above a defective enzyme is present; fibrin is not produced. These rare individuals suffer from a condition known as **haemophilia,** which is a sex-linked characteristic inherited from one of their parents (see page 175). Haemophiliacs can now be treated with drugs that enable their blood to clot. In the past they usually died of severe blood loss following a simple injury. Female haemophiliacs invariably died at the onset of their first menstruation.

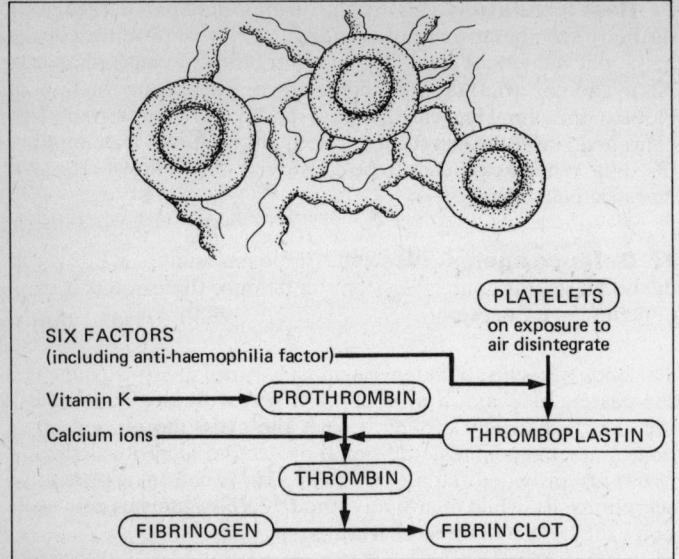

Figure 28. Blood clot and clotting process

Inflammation. Micro-organisms can harm the body in two ways: directly, by digesting cells, or indirectly, by producing enzymes or waste products (**toxins**) that have a widespread damaging effect on the working of the body's cells.

The blood is able to defend the body against micro-organisms and/or their toxins. Following an infection one of the first things to happen is a localized dilation of the blood vessels near the site of the entry of the causative agents. This has two effects. Firstly, it increases the concentration of phagocytic leucocytes in the infected area. These cells are able to engulf the invading micro-organisms. Secondly, there is an increase in the local concentration of antibodies, which are able to adhere to the micro-organisms and/or their toxins thereby neutralizing them.

At about the same time there occurs an increase in the permeability of the dilated blood vessels, causing a greater flow of blood into the tissue fluid surrounding the infection. This produces a localized swelling, the source of the natural redness and soreness of a cut or wound. In a confined area, such as a knee or elbow joint, this swelling can be very painful.

The resulting mixture of leucocytes (many of them dead as a result of their efforts) and damaged tissues forms a white substance **pus,** which is gradually reabsorbed by the body as the infection subsides.

Most infections are checked by this localized increase in blood flow and the plugging of a wound by a blood clot. However, certain micro-organisms invade the body more quickly than others. Some bacteria are even protected from the action of leucocytes by being enclosed in a capsule.

In response to such life-threatening micro-organisms the whole of the body's metabolism is stepped up as more and more antibodies and leucocytes are produced. This increase in tissue activity produces the rise in body temperature characteristic of an illness.

The body's natural defence mechanisms give rise to either an **acute** or a **chronic** response to an invading foreign substance or organism. An acute response is short-lived and results in either complete recovery or death. A chronic response is longer lasting and occurs when the causative agent cannot be overcome readily. With micro-organisms such as bacteria, usually they remain localized at the site of infection but from time to time they or their toxins escape and reinfect the body. An example of such a recurrent disease is tuberculosis (TB) (see page 185).

Following recovery from an infection, antibodies may remain in the blood for a long time, providing protection against a possible reinvasion by the same micro-organism. This form of protection from infection is known as **immunity.** Immunity can be natural, that is produced by the body as a result of a previous infection, or acquired artificially by means of an injection with a vaccine (see Chapter 9).

Blood transfusions

The body of an average-sized adult contains about 5.5 litres of blood, although the volume changes from time to time. If more than 40 per cent of this is lost over a short period of time, the body cannot make new blood quickly enough before death occurs. But this eventuality can be prevented by replacing lost blood with that from another person, called a donor, by means of a transfusion. However, because blood varies slightly from person to person, donated blood cannot be given to anyone. It has first to be classified according to its blood group and so checked for compatibility with the blood of the person receiving the transfusion, the recipient.

Blood groups. Blood can be classified according to a number of different systems. The most commonly used and most important is known as the **ABO system.** This was discovered in 1900 by Karl Landsteiner, working in Vienna. Until Landsteiner's discovery, blood transfusion was extremely hazardous because if a patient was given blood of the wrong ABO group his own blood would cause clumping, and might destroy, the donated cells, forming potentially lethal blood clots in small vessels.

Blood groups are based on the types of **antigens** (special proteins) found on the surface of the red blood cells. In the ABO system, if antigen A is present on the cells, then that blood is referred to as blood group A. Group B blood contains red cells with antigen B on their surface. Group AB blood contains red cells with both antigen A and antigen B. The red cells of O blood group possess neither of these antigens.

A and B antigens have complementary antibodies present in the plasma: these are antibody A and antibody B (usually written as anti-A and anti-B). Plasma containing anti-A will cause clumping of red cells containing antigen A and plasma with anti-B will similarly effect antigen B cells. Blood does not normally clot because blood group A individuals possess plasma with only anti-B and blood group B individuals have only anti-A in their plasma. In individuals of blood group AB neither anti-A nor anti-B are present in the plasma, whereas in group O individuals both anti-A and anti-B exist. (See chart opposite.)

A clot, indicated by a minus sign, will occur if donor blood contains an antigen that matches an antibody in the recipient's blood. Thus blood of group A given to a person with blood of group B will clot because the antigen A of the donor's red cells will react with anti-A in the plasma of the recipient.

Because blood group O contains no antigens, it can be safely given to anyone. It is sometimes referred to as the **"universal donor"** blood group. However blood group O individuals can only *receive* blood from blood group O donors; this is because the anti-A and anti-B in their plasma will cause clotting of any donor red cells that contain either A or B. (In a transfusion in which O group blood is donated, the anti-A and anti-B present become so greatly diluted as to have no effect.)

	Recipient's Blood			
Donor Blood	O ab	A b	B a	AB
O ab	+	+	+	+
A b	−	+	−	+
B a	−	−	+	+
AB	−	−	−	+

The capital letters refer to the blood group, the small letters to the antibodies in the plasma.

AB individuals can receive blood from anyone; their plasma lacks any antibodies that could react with donated red cells. They are therefore referred to as **"universal recipients"**.

The Rhesus system. One other blood group system also needs to be considered before a transfusion is carried out. This is known as the **Rhesus system,** so-called because it was first discovered by injecting rabbits with red cells obtained from a Rhesus monkey.

About 85 per cent of the UK population have the Rhesus antigen on their red blood cells. These people are known as Rhesus positive (Rh+). The remainder lack the antigen and are called Rhesus negative (Rh−). Unlike the ABO system, Rh^- blood does not necessarily carry the Rh antibody in its plasma. However, if Rh^+ blood finds its way into a person with Rh^- blood, the Rh antigen will stimulate the recipient to produce the Rh antibody. Nothing further happens, but if at a later date the recipient receives more Rh^+ blood, the Rh antibodies now in its plasma will cause clotting of the donated Rh^+ red cells. There is little chance of this sequence of events happening except during pregnancy.

Towards the end of a pregnancy, there is a strong possibility that a small quantity of foetal red blood cells will leak across the placenta into the mother's bloodstream. This can prove dangerous if a Rh^- negative woman is carrying a Rh^+ foetus. (How this Rh incompatibility arises is explained in Chapter 8). The leaked Rh^+ blood will stimulate the mother to produce Rh antibodies. This will not have any effect for the moment, although some of the antibody can leak back into the foetal circulation causing some red cell destruction. (Usually the child is born before this can happen.) But if during a subsequent pregnancy the Rh^- mother has a second Rh^+ foetus, the Rh antibodies now present in her blood will leak across the placenta and destroy many of the foetus's red blood cells. This condition is known as haemolytic disease of the newborn **(erythroblastosis foetalis)** and may be fatal if the child's blood is not immediately transfused with the Rh^- blood. The transfusion can also be carried out while the child is still in the womb.

In the past, Rh^- women were therefore able to have only one or, if fortunate, two children, if her husband was Rh^+. But now haemolytic disease can be prevented by treating the mother with an anti-Rhesus protein that coats any foetal red cells that escape into her circulation and thereby prevent her producing any Rh antibodies.

The blood circulatory system

Blood is circulated to all parts of the body, so that the needs of each and every cell can be constantly met. Should the blood circulatory system stop for any reason, the consequences are considerable. Within 3 to 5 seconds the brain loses consciousness. Some 10 seconds later the body starts to twitch convulsively. After 9 minutes without circulating blood, the mental powers of the brain are irreversibly affected. Death follows soon after. The heart, too, cannot survive for long without fresh circulating blood. After 30 minutes it becomes irreparably damaged, and will not beat again.

The structure of the heart. The heart is situated between the lungs and directly behind the sternum, almost in the centre of the body. It is encased by a thin membrane, the **pericardium.** Between the pericardium and heart is a thin layer of fluid which prevents friction as the heart beats.

The heart consists of two pumps that effectively have been wielded together. Each pump has two compartments: an antechamber or **atrium** (sometimes called an auricle), and a **ventricle.** Ventricles have very thick muscular walls, four times thicker than the walls of an atrium. The left ventricle has much thicker walls than the right ventricle. The two pumps are fused together by an even thicker dividing wall called the **septum.**

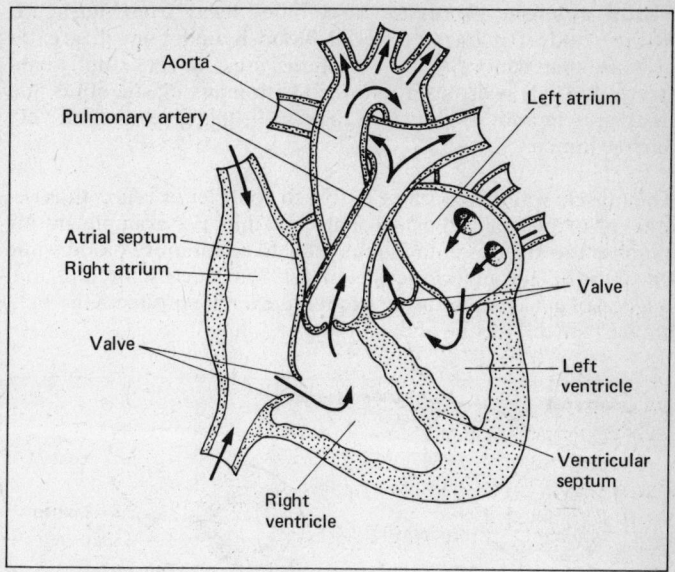

Figure 29. Structure of the heart (cut in section)

The atria receive blood returning *to* the heart from the general circulation, the left atrium receiving blood from the lungs and the right atrium receiving blood from the rest of the body. The ventricles propel blood *away from* the heart into the general circulation, the right ventricle pumping blood to the lungs and and the left ventricle pumping blood to the rest of the body. Valves between the heart chambers and at the exit and entrance of the main blood vessels control the direction of blood flow.

Spiral muscle fibres wind round both ventricles in the form of a common envelope, squeezing or wringing out the blood at each contraction. In addition, the left ventricle has extra spiral fibres.

The two atria are also able to contract, pumping blood into their respective ventricles. But usually this action is not necessary as the elastic recoil of the ventricles following contraction draws blood into the ventricles through the atrial-ventricular valves. (Atrial contraction would become important should these valves function incorrectly through disease.)

Blood vessels. Arteries carry blood away from the heart, **veins** conduct it back. Arterial blood is under much greater pressure than venous blood, so arteries must be very stoutly constructed. Their walls are reinforced with many elastic fibres and thick muscle and, as shown in Figure 30, this gives them a very narrow lumen.

The muscle wall of an artery is able to contract or relax, thus decreasing or increasing the flow of blood within. For example, during exercise the arteries in the limbs relax to admit more blood while those to the gut and kidneys contract. Particularly forceful contraction of gut arteries during strenuous exercise produces the well-known "stitch" feeling of pain in the abdomen.

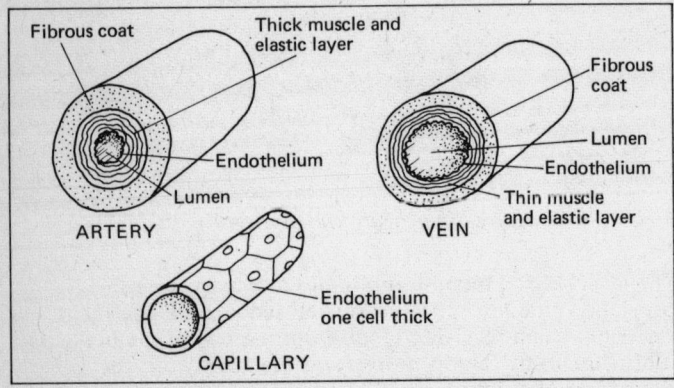

Figure 30. Structure of an artery, vein and capillary

With the exception of the pulmonary arteries that carry deoxygenated blood to the lungs for gaseous exchange (see page 64), all of the body's arteries carry oxygenated blood and originate from the **aorta,** which is the huge artery leaving the heart from the left ventricle. The aorta is about 2 cm in diameter, and as it descends through the body it divides like a branch of a tree, giving off several pairs of arteries. Its main branches are the coronary arteries, which supply the muscles of the heart itself, the carotids, which run up the neck and serve the head, the subclavians serving the arms, the coeliac and mesenteric arteries, which supply the gut, the renals serving the kidneys, and two iliac arteries which serve the legs (see Figure 25).

Branches of arteries are often linked together. This means that if one section of an artery becomes blocked for any reason, an alternative route for the blood exists (see page 89).

Arteries are protected from physical damage by being located deep inside the body. Should one be cut, however, a jet of blood will spurt out with great force in tune with the beating of the heart. Severe bleeding can be stopped by applying pressure above and below the cut ends.

As arteries reach further into the tissues they become smaller and thinner walled and are known as **arterioles.** Arterioles eventually lead into a network of microscopic vessels, the **capillaries.** It is at the capillaries that the blood is able to transfer its cargo of oxygen and food etc. to the tissues and pick up waste products.

From the capillaries, blood is returned to the heart via the venous blood system. This consists of small thin-walled tubes called **venules** that eventually connect up to form larger vessels, the veins. The high pressure of blood in the arteries (about 120 mm Hg) is expended by the time the blood has passed through the capillaries and reached the veins. Therefore veins have much thinner walls than arteries, but their diameter is greater to allow the blood to flow with less resistance.

	AORTA	ARTERY	ARTERIOLE	CAPILLARY	VENULE	VEIN	VENA CAVA
DIAMETER	2 cm	4 mm	30μm	8μm	40μm	5 mm	3 cm
WALL THICKNESS	2 mm	1 mm	20μm	1μm	2μm	.5 mm	1.5 mm

Figure 31. Dimensions of blood vessels

The veins can contract and relax a little to accommodate a variable return of blood to the heart. Also to prevent backflow of blood, they contain pairs of valves along their length. The flow of blood in veins is sustained by the continuous movement of muscles all over the body, which gently squeezes together the walls of the vessels.

Contraction of the heart. Heart muscle has an inherent ability to contract rhythmically. Left to their own devices, the atria will contract about 140 times a minute and the ventricles about 30 times a minute. Heart muscle cells are different from other muscle cells in that they are branched to form a network and there are no distinct boundaries between fibres or cells (see page 22). They act as one cell. So when heart muscle begins to contract, the whole heart is affected. This has many advantages, the most important being that the heart is able to generate enough co-ordinated force to jetison the blood out of both ventricles almost simultaneously.

In the wall of the right atrium is a tiny knot of very active heart muscle that is richly supplied with nerves. This area is called the **sinoatrial node** (or pacemaker) and is the source of the heart beat. A wave of nervous excitation begins here and spreads across the two atria, causing them to contract together. While the atria contract, the ventricles relax in order to receive the next volume of blood. When the wave of contraction reaches the atrial-ventricular junction, it is transferred to the ventricles by a bundle of connecting fibres, the bundle of His. Contraction of the ventricles is called **systole** and the filling, **diastole.**

The pacemaker's inherent rate of nerve impulse transmission produces the heart's natural rate of contraction described above. But this is constantly overriden by a continuous stream of impulses which reach the pacemaker along the vagus nerve from the brain. The vagus is part of the parasympathetic nervous system (see page 114), which inhibits muscle action, and it is this which directly controls the rate at which the heart beats. For the heart rate to increase, the number of impulses passing along the vagus nerve must be reduced.

Effect of exercise. The heart increases its rate of contraction in order to meet additional demands from tissues for oxygen and food. At rest it contracts (beats) about 70 times a minute, each beat despatching 70 mm^3 of blood from each ventricle (the

stroke volume). Thus the volume of blood pumped from one ventricle, the **cardiac output,** is almost 5 litres of blood per minute, equivalent to the total body blood volume. Looked at another way, when the body is at rest a single erythrocyte takes about 1 minute to travel from the heart, through the circulation, and back again.

During strenuous exertion, cardiac output may increase by as much as eight times. This is achieved by both an increase in the stroke volume – up to 200 mm^3 – and an increase in heart-rate from 70 to 200 beats per minute.

Heart-rate is also increased by the action of the hormone adrenaline (see page 119).

Blood pressure. The pressure at which blood is pumped into the arteries varies considerably between individuals; the average is about 120 mm Hg. But a person's blood pressure fluctuates during the beat itself, from 120 mm Hg during systole to 80 mm Hg during diastole. The high arterial pressure is maintained as the blood flows unimpeded through the aorta and its main branches but drops as it flows through the arterioles. By the time blood reaches the capillaries it has dropped to 32 mm Hg.

The function of the blood pressure is to maintain the flow of blood to the vital organs. Insufficient pressure delivers insufficient blood, while excessive pressure places extra strain on the heart and blood vessels. At rest and during exercise blood pressure should remain fairly constant. As they get older, many people experience a fairly permanent increase in blood pressure known as hypertension. This can be due to an increased resistance to the flow of blood through a diseased or failing organ such as the kidney, but more commonly it has no recognized cause; it is then known as **essential hypertension.**

Essential hypertension can be asymptomatic, that is produce no **symptoms,** but it can aggravate other disorders, such as arterial disease, or injure the kidneys. High blood pressure certainly reduces the chances of a person living to a ripe old age. The condition is generally treated by drugs that minimize or prevent the contraction of arterioles so that their resistance to the flow of blood is reduced. Recently it has been shown that an increase in dietary potassium and a decrease in dietary sodium will reduce the blood pressure of a hypertensive patient almost as well.

Defects of the heart and circulation

The sole function of the heart is to pump blood round the circulatory system. It can thus have only one major defect, namely failure to pump. But although the heart rarely fails completely, many minor defects can occur that reduce its efficiency. These can be detected by medical examination of the pulse, the blood pressure and the sounds of the heart.

The **pulse** corresponds to the rate at which the heart beats. It can be detected in the neck or the wrist by applying light pressure to an artery. Blood pressure can be determined by a device called a **sphygmomanometer.** This consists of an inflatable cuff which is connected to a mercury manometer. The cuff is applied to the arm and inflated until it produces sufficient pressure to stop the flow of blood in the brachial artery. Air is then released from the cuff by means of a valve until the pulse returns. At this point the level of mercury in the manometer corresponds to the systolic pressure. A stethoscope is used to listen for the return of the pulse in the wrist (see Figure 32).

The abrupt closing of the valves in the heart can be heard as murmurs by applying a stethoscope to the chest wall. A trained ear can assess the loudness of the murmurs and also whether they are followed by any abnormal sounds, which would indicate a heart valve defect. Typically, the abnormal murmurs resemble the sigh of a gust of wind leaking through a broken window pane; they are produced by blood flowing through a narrowed or incompletely shut valve. A leaky valve can reduce the efficiency of the heart greatly and may need to be replaced by a synthetic one.

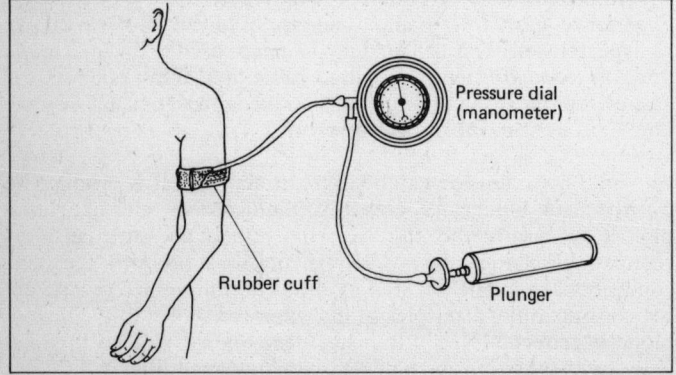

Figure 32. Use of a sphygmomanometer

Other heart defects follow from disease of the coronary arteries. These tiny vessels supply blood to the actual muscle of the heart. Coronary artery disease begins with fatty deposits called plaques collecting in the lumen of an artery. As the deposits build up they impede the flow of blood. Eventually they can block the artery completely, thereby stopping the flow of blood to the corresponding part of the heart muscle, which then dies. This produces a **heart attack,** and is *very* painful. Contraction of the heart becomes impeded and may stop altogether.

A similar process can occur in the arteries or arterioles supplying the brain. The result is death of the part of the brain where the blockage occurs, known as a **stroke.** The effect of the stroke is variable and depends upon how much and which part of the brain is destroyed.

The most common heart defect is that involving the pacemaker. For some reason it loses its ability to generate impulses. It can be rectified by the means of an artificial pacemaker, which is implanted beneath the shoulder and connected by an electrode to the heart. This device delivers a regular impulse to the heart and substitutes for the diseased pacemaker.

A diseased heart can be replaced by a healthy one taken from another person who perhaps has died in a road accident. This operation was first carried out by the South African heart surgeon Dr. Christiaan Barnard in 1967. Since then hundreds of similar **transplants** have been carried out in various hospitals throughout the world. On the whole these operations have not been very successful because of the problems associated with **rejection** of the donated heart (see page 182). At two U.S. hospitals (Texas and Utah) attempts have also been made to replace a living heart with a plastic equivalent. The artificial pump is worked by compressed air that is generated by an externally worn motor and piped into the heart through tubes passing under the ribs. As yet they are only experimental, but in years to come they may be of benefit to thousands of patients who would otherwise die.

Failure in the functioning of an artery accounts for about a third of all deaths in the UK; stroke and coronary heart disease account for most of these.

Diseased arteries can be repaired by surgery or even replaced with synthetic tubing. An example of this is the Dardik graft, which can be used to replace an artery in the leg. It is particularly interesting because the graft material originates from veins obtained from the discarded umbilical cords of American babies. The umbilical veins are treated with chemicals that harden them and make them safe to transplant in a patient's leg.

The lymphatic system

The lymphatic system returns excess tissue fluid to the bloodstream and plays a major role in the body's defence against infection (see Chapter 9). Lymph capillaries, narrow thin-walled tubes, drain mainly protein molecules from the tissues. Lymph fluid then drains into larger lymph vessels.

These lymph vessels lie between muscles, so movement of lymph occurs with muscle contraction, the direction of flow being controlled by valves situated inside the vessels. The main lymph vessels unite to form two major lymphatic ducts, which empty the lymph into the subclavian veins below the neck. In certain places lymph vessels enlarge to form lymph nodes or glands; these are particularly prominent in the neck, upper thorax and groin. (The tonsils and the spleen are lymph nodes.) They defend the body against infection by:
1. manufacturing lymphocytes, which are responsible for antibody production, and
2. filtering lymph as it passes through the nodes, enabling phagocytes in these areas to ingest any foreign bodies.

Key terms

Antibody Protein produced by leucocytes on contact with a substance or invading micro-organism that acts as an antigen (ANTI-foreign BODY); antibodies neutralize the antigen and thus have a protective function.
Antigen Any substance or organism that is treated as foreign by the body and stimulates antibody production (i.e. GENerates ANTIbodies).
Arteriole Small blood vessel, between an artery and a capillary.
Artery Tough, muscular, thick-walled vessel that carries blood away from the heart.
Capillary Thin-walled vessel that allows substances to pass from blood to the tissues.
Diastole Relaxation of the ventricles of the heart.
Erythrocyte Red blood cell.
Haemophilia A disease in which the blood will not clot.

Heart attack Failure of the heart to pump, usually caused by an interruption to the flow of blood to the heart muscle via the coronary circulation.
Hormones Chemical substances secreted into the bloodstream that produce an effect on a target organ.
Hypertension An abnormally raised blood pressure.
Inflammation A defensive reaction by the body to an injury or infection, causing swelling and redness. It promotes healing.
Leucocyte White blood cell.
Lymph Straw-coloured liquid similar to plasma derived from tissue fluid.
Lumen Central cavity of a tube or vessel.
Plasma Blood without its cells and platelets.
Platelets Small fragments of cells that on exposure to air will cause the blood to clot.
Pacemaker Small section of the right atrium which generates the nervous impulse that causes the heart to contract rhythmically.
Pulse The heart beat as is felt in the wrist or the neck.
Sphygmomanometer Device that measures the pressure of blood.
Stethoscope Instrument that allows the sounds of the body's heart and lungs to be heard.
Stroke An interruption in blood flow to the brain or part of it.
Systole Contraction of the ventricles of the heart.
Tissue fluid A fluid that bathes the cells of the body, derived from plasma.
Transfusion The transfer of blood from a healthy person to a sick person.
Vascular system A system of tubes or vessels.
Vein Carries blood to the heart; thin-walled but wider lumen than arteries.
Venule Small blood vessel linking capillaries to veins.

Chapter 5
The Kidneys, Skin and Excretion

Excretion is the process by which a living organism removes metabolic waste products from the body. The waste products include carbon dioxide, nitrogenous substances such as urea, and excess water and salt. All of them are toxic if they remain inside the body. They originate in the body's cells. They pass into the blood and are carried to one of a number of excretory organs for removal to the external environment. The principal excretory organ is the kidney. Others include the lungs, the skin and the liver.

Excretion should not be confused with **defaecation** (also called egestion), which is the removal of undigested food from the gut. Remember that the inside of the gut is not part of the internal environment of an organism (see page 34). Undigested food that is passed out of the gut as faeces has never entered the body's cells and so is not a waste product of metabolism.

The kidneys

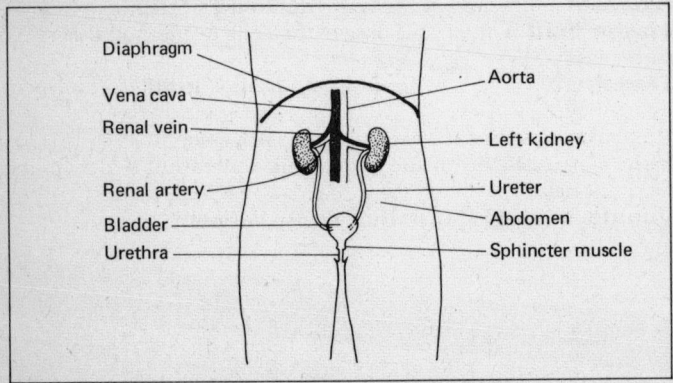

Figure 33. Position of the kidneys

All mammals have two kidneys. These are found on either side of the vertebral column at the back of the abdominal cavity. Their job is to filter metabolic wastes from the blood and maintain the blood's osmotic pressure (see page 27).

The kidneys are supplied with blood by the renal arteries, one going to each kidney. After filtration, blood returns to the heart via the renal veins. Within the kidneys, blood vessels run alongside long tiny tubes called **nephrons;** each kidney contains about one million of these. At one end of a nephron is a swollen cup-like structure, the **Bowman's capsule.** The other end runs into a funnel-like structure, the pelvis, the narrow part of which becomes a tube called the **ureter.**

The Bowman's capsules are all located in the noticeably darker outer region of each kidney, the **cortex.** The greater part of each nephron is found in the lighter region, which is called the **medulla.**

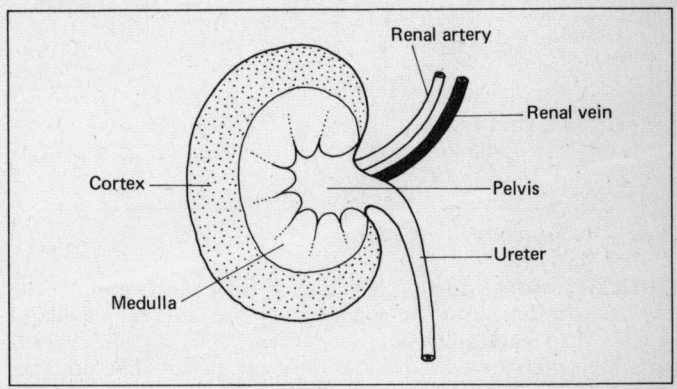

Figure 34. Structure of a kidney

Each Bowman's capsule is rather like a football which has been pushed in at one side. The cup thus formed is filled with a dense knot of capillaries, the **glomerulus,** such that the outer wall of the capsule and the walls of the capillaries are closely associated. In fact, only two layers of cells, one capillary and one capsular, separate the blood from the capsule interior.

Leading from the capsule is the **proximal convoluted tubule,** which in turn leads to the **loop of Henlé.** This plunges down into the medulla of the kidney then leads back up to the cortex. Here it is continuous with the **distal convoluted tubule,** which feeds into a collecting duct. Each collecting duct serves many nephrons and leads down to the pelvis of the kidney. This, in turn, releases its contents into the ureter. From both kidneys, the ureters conduct urine to the bladder, where it is stored before passing out of the body.

Functioning of the kidneys. The kidneys' job of filtering the blood is achieved through two main processes – ultrafiltration and selective reabsorption.

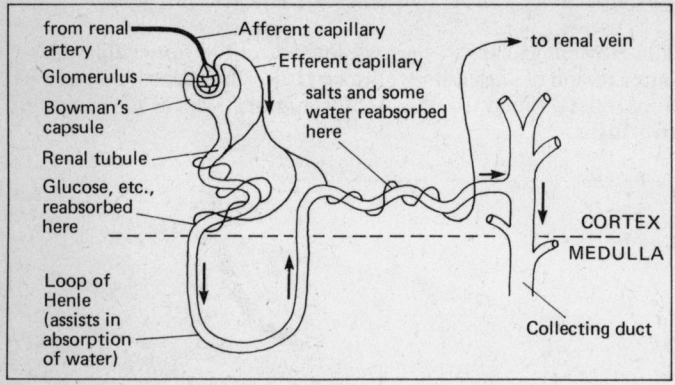

Figure 35. Structure of a nephron

Ultrafiltration is carried out in each Bowman's capsule. The capillary leading into the glomerulus (the afferent capillary) is wider than that leading out (the efferent capillary), and this sets up a high pressure within the knot of capillaries. The pressure causes small molecules in the blood, such as glucose, amino acids, salts, water and urea, to filter through the capillary walls into the cavity of the Bowman's capsule. Blood cells and protein molecules such as fibrinogen remain in the capillaries as they are too large to be filtered. Ultrafiltration is a passive process; it does not require the expenditure of energy.

If all substances passing into the Bowman's capsule (together forming the glomerular filtrate) were to form urine, many valuable chemicals would be lost from the body. In the nehpron tubules a process known as **selective reabsorbption** occurs in which some substances in the filtrate are reabsorbed back into adjoining blood capillaries.

In the proximal tubule, glucose and some water are reabsorbed into the capillaries that envelope this structure. In the distal tubule, more water, together with sodium, potassium and chloride ions, are reabsorbed into the blood. Water is also reabsorbed in the loop of Henlé. Most of this reabsorption occurs by **active transport** (see page 28).

Urine. The end-product of ultrafiltration and selective reabsorption is urine, which contains water, urea and uric acid, salts, traces of hormones, water-soluble vitamins and any drugs the person may be taking. If the blood is too dilute – for example, after drinking a large amount of fluids – less water is reabsorbed from the kidney tubules than normal and the urine is dilute. If the blood is too concentrated – for example, after profuse sweating – more water is reabsorbed than normal and a concentrated urine is produced.

Urine drains into the **bladder.** This is a muscular sac located at the base of the abdomen. At its base is a ring of muscle, a sphincter, which controls the release of urine, the process called **micturition.** The sphincter comes under conscious control early in life (during potty training) so that urine is passed only at appropriate times. As the bladder fills with urine, its walls stretch. This stimulates stretch receptors in the bladder wall that inform the brain about how much urine is present.

Regulation of salts and water by the kidney

The excretory role of the kidneys is straightforward: urea is totally removed from the blood, along with other toxic materials. Its other role, that of an osmoregulator, is rather more complex.

Receptor cells in the **hypothalamus** at the base of the pituitary gland (see page 111) continuously monitor the **osmotic pressure** of the blood. They are stimulated to increase their output of nerve impulses when the osmotic pressure becomes too high. The impulses trigger cells in the posterior lobe of the pituitary to release **antidiuretic hormone** (ADH). ADH stimulates the kidneys to produce smaller quantities of, or more concentrated, urine by causing an increase in water reabsorption from the distal tubules and collecting ducts.

As more water is reabsorbed, the blood becomes less concentrated. This in turn lowers the blood's osmotic pressure so the osmoreceptor cells send fewer signals to the pituitary's ADH secreting cells. As less ADH is produced, filtration in the kidneys involves less water reabsorption, and so on. This mechanism is an example of a homeostatic **negative feedback** process – the body maintaining the internal environment at an optimal point (see page 102).

NOTE. **Diuretics** are drugs sometimes prescribed by doctors that cause patients to produce copious amounts of dilute urine by having the opposite effect to ADH. They are used to reduce blood volume in conditions such as congestive heart failure where the subject has too much fluid for the heart to cope with.

The lungs and excretion

The lungs can also be regarded as organs of excretion for it is via the lungs that carbon dioxide is removed as well as some water. Carbon dioxide is a major waste product of respiration. During breathing, it can be shown that inhaled air contains relatively little carbon dioxide – 0.4 per cent – but in exhaled air the proportion is tenfold higher – 4 per cent (see page 69).

The carbon dioxide is carried from the tissues to the lungs in blood plasma. The gas diffuses from cells into the plasma and combines with water to form carbonic acid H_2CO_3, a reaction involving the enzyme carbonic anhydrase. The acid then dissociates into bicarbonate and hydrogen ions until the plasma reaches the lungs (see page 76). Also a tiny amount of the carbon dioxide combines with the haemoglobin of red blood cells.

When the blood reaches the capillaries around the alveoli, the low carbon dioxide concentration in the inhaled air triggers a reversal of the process. The carbonic acid is converted to carbon dioxide and water and the carbon dioxide diffuses into the alveoli. The gas is then expelled during exhalation.

The skin and temperature regulation

The skin is the major organ of temperature regulation in that it controls the amount of heat lost to or gained from the external environment.

In general, the body temperature of man remains at around 37°C irrespective of fluctuations in the external environment. This is an optimum temperature for the functioning of enzymes that control metabolism.

If body temperature rises above 42°C death usually occurs. Above about 40°C proteins become denatured, and since all enzymes are proteins, at 42°C these cease to work and the body's metabolism grinds to a halt. Similarly, if body temperature falls too far the rate of enzyme activity drops and the metabolic rate slows right down. Again, eventually death can occur. This is why so-called "exposure" suffered by mountaineers and hikers is taken so seriously by doctors. Below a certain level, body metabolism cannot generate enough heat to compensate for the loss via the skin and lungs.

With regard to temperature control, animals can be classified into one of two categories. **Homoiothermic** animals, such as man, generate heat from their metabolism and regulate heat loss to maintain a constant internal temperature. **Poikilothermic animals,** such as lizards and snakes, are not so advanced physiologically, and their body temperature is dependent on the environmental temperature.

But this classification can be misleading for poikilotherms are often referred to as cold-blooded animals even though frequently their blood is not at all cold. In fact in some cases the blood temperature of poikilotherms can be well above that of the environment.

A better classification is between **endothermic** and **ectothermic** animals. As the words imply, endotherms generate heat internally and maintain their temperature by physiological mechanisms whereas ectotherms gain heat from the environment and maintain a relatively constant body temperature by behavioural mechanisms, e.g. crocodiles sunbathe to gain heat or retreat to the water when they become too hot. In endotherms such as man heat is produced by metabolic activity.

The body can **lose** heat by one of four mechanisms:

(a) Radiation. Heat is a form of energy and can therefore be radiated from the body as electromagnetic radiation. A man in a room at 21°C can lose up to 60 per cent of his heat in this way.

(b) Evaporation. Changing a liquid into a vapour requires heat. Thus sweat released on to the skin surface acts as a major coolant as a result of its evaporation. However, its efficiency is dependent on environmental factors such as air humidity, air temperature and the presence and degree of air currents.

(c) Convection. Air close to the body is warmed by direct contact and by radiation. Because pockets of warm air rise, a micro-circulation of air is created around the body. (This may further aid evaporation.)

(d) Conduction. This is the transfer of heat from a hot body to a cool body in physical contact with it. Walking barefoot on a cold stone floor will result in rapid cooling of the feet as heat is lost to the floor. Similarly a hot beach may be unbearable to walk on barefoot because of the acute heat transfer to the soles of the feet.

Structure of the skin. The structure of mammalian skin is the key to its success as a thermal regulator (Figure 36).

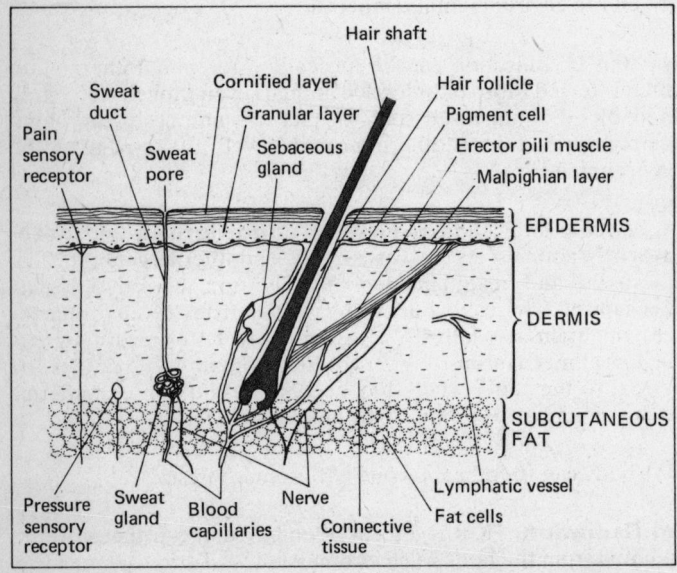

Figure 36. Section through human skin

The skin is a multi-layered organ. At the base is the subcutaneous fat layer, which is more or less developed according to age, race, sex and nutritional status. Above this, and with a variety of structures embedded in it, is a thick layer of connective tissue containing many elastic fibres. This is the **dermis.** Its main structures are capillary nets, hair follicles, sebaceous glands, sweat glands and heat and cold receptors that are linked to the brain by nerves.

Above the dermis is the **epidermis.** This also is multi-layered. At its base is the actively dividing **malpighian layer.** At the surface of the epidermis skin cells are constantly being worn away only to be replaced from the malpighian layer. The **sweat ducts** and hair follicles penetrate through the epidermis, giving access to the skin surface.

Hair is an important part of the skin. It is made in the hair follicles and is composed of the protein **keratin.** Hair contains variable amounts of the pigment **melanin** that gives it its colour. Hence people with brown hair have more melanin in their hair than blonde people. The amount of melanin incorporated into hair is genetically inherited (see Chapter 8).

Production of hair is a continuous process and even carries on for a short time after death. In some animals the rates of shedding and growth of hair vary with the seasons, as with the winter and summer coats of dogs and cats. This variation may also relate to melanin levels; some arctic animals have a brown summer coat and a white winter coat, this being vital for camouflage.

Response to cold by an endotherm. When an endotherm loses heat too rapidly the blood begins to cool. Blood temperature is monitored by the thermoregulatory centre in the hypothalamus of the brain. In response to a fall in blood temperature this transmits impulses to effectors such as the erector pili muscles and sweat glands in the skin. Their actions and those of other regulatory mechanisms restore normality.

1. **The hair is raised** to a more or less vertical position by contraction of the erector pili muscles (see Figure 36). This traps air in the spaces between the hairs. The air becomes warmed by the body and acts as an insulator. In man this process is not very efficient because of the hair's shortness but it is aided by the wearing of clothes.

2. **Superficial blood vessels** in the skin contract (vasoconstriction). This means less blood reaches the surface of the skin and, since the blood carries a lot of body heat, less heat is lost to the environment. In exposed parts of the body such as the ear-

lobes shunt vessels exist between arterioles and venules in the skin that in cold conditions expand to divert blood from the capillaries altogether. In addition, some blood is taken up into major blood reservoirs such as the liver and spleen. This reduces the total volume of blood circulating, thereby helping to conserve heat.

For these reasons extremeties may appear bluish in cold weather, for the reddening effect of blood circulating near the surface is reduced. In extreme circumstances of prolonged cold, cells in the peripheral areas can be denied food and oxygen for so long they die. This condition is known as frost-bite.

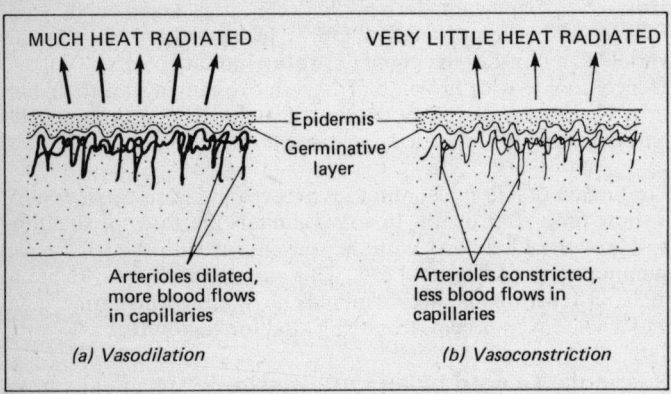

Figure 37. Diagram to show action of shunt vessels

3. **The body produces extra heat** by increasing the metabolic rate in the liver and muscles. This is marked by a general stiffening of the muscles but can progress to spasmodic contraction of skeletal muscle (shivering) and contraction of smooth muscle in the skin (producing goose-pimples). In addition, fatty deposits around the body, collectively termed **brown fat,** "burn" sugars to release heat and so raise blood temperature. In Eskimos this has been found to be a particularly well-developed mechanism.

4. **The presence of subcutaneous fat,** a passive insulating structure, reduces heat loss. In animals adapted to cold climates this layer of skin can be very well developed, as represented by the blubber of whales, seals and other polar creatures.

5. **Behavioural mechanisms.** It should also be remembered that man often enhances heat production by indulging in vigorous

exercise. A commonly observed example of this is at bus-stops on cold days where would-be passengers can be seen stamping their feet, slapping their arms and jumping up and down in order to generate body heat.

Response to heat by an endotherm. In hot environments the processes outlined above are put into reverse. Heat production is minimized and heat loss encouraged.

1. **Hair is lowered** by relaxation of the erector pili muscles so that it lies flat and ceases to form an insulating layer.

NOTE: When the environmental temperature is higher than body temperature, heat gain from the surroundings may become a hazard. At this stage the hairs will be raised again, forming an insulating layer of relatively cool air. Heat uptake is then slowed or prevented. This is vital to large desert animals such as camels that cannot easily escape from the sun.

2. **Superficial blood vessels in the skin dilate** (vasodilation). This brings blood close to the skin surface to increase heat loss. Where shunt vessels are present, they are constricted to ensure the blood goes through the capillaries. Total blood volume is increased by closing down the major reservoirs, thereby raising the flow of blood to the skin.

3. **Sweating.** This involves the secretion of a watery fluid from the sweat glands on to the skin surface. As the fluid evaporates, the skin and the blood flowing through it are cooled. It is a very important regulatory mechanism because when environmental temperature exceeds body temperature there is no other way of losing heat. But it is not without disadvantages. If air humidity is low then sweat evaporation and, hence, cooling, occurs rapidly. However, if humidity is high then evaporation is limited by the remaining capacity of the air to take up more water vapour. So in the dry air of the Sahara, for example, sweating is an efficient way of losing heat, but in the humid areas of West Africa it is much less efficient. Assuming air temperature is the same in both places, on a hot day a man may be able to work quite hard in North Africa but in West Africa he would collapse with heat stroke if asked to perform the same task.

Sweating involves water loss and, to a limited extent, salt loss. These both come from the blood. The sweating mechanism therefore alters blood composition, and this is reflected in a corresponding alteration of kidney function. Copious sweating results in reduced urine production.

In animals with few or no sweat glands (such as cats and dogs), heat is lost by panting. Evaporation takes place from the lung surface.

4. **The metabolic rate falls** – this occurs in hot weather so that less heat is generated internally. This is why animals are generally less active in hot weather than in cold weather.

5. There is generally **less subcutaneous fat** in animals that normally inhabit warm environments. Where fat storage is necessary for reasons other than insulation, such as inconsistent food sources, the fat is localized so that it does not interfere with heat loss. In camels, fat is stored in their humps.

6. When it is hot, animals frequently **seek shade.** Where daytime temperatures are always very high many species are nocturnal, thereby avoiding overheating and desiccation.

The processes described above all form part of the homeostatic mechanism that controls body temperature (see Figure 38).

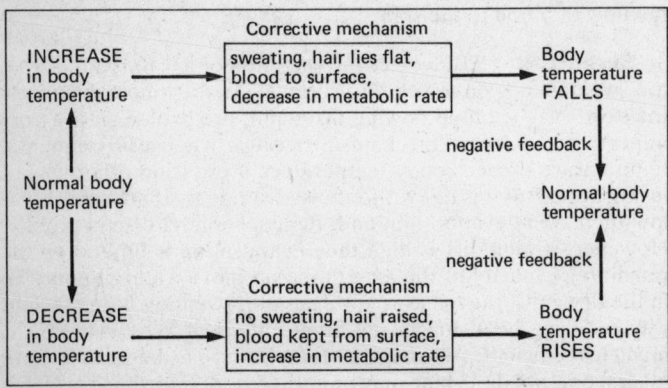

Figure 38. Homeostatic control of body temperature

Other functions of the skin

Physiologically the skin acts as a barrier to excessive water loss from the body and is therefore a vital adjunct to osmoregulation in terrestrial animals. In addition it protects the internal environment from the sun's radiation and from mechanical damage and acts as a physical barrier to the entrance of foreign bodies and pathogenic organisms.

The sebaceous glands open into the hair follicles and produce an oily secretion called **sebum.** This is slightly antiseptic and so helps in preventing bacterial growth on the skin. It also makes the hairs water-repellant and keeps the epidermis supple so that it is less prone to cracking. A cut or a tiny crack in the epidermis can soon become infected by micro-organisms.

In the malpighian layer are the melanocytes. These cells, like hair, contain the pigment melanin. They are more numerous in the skin of people of the Negro race, giving such individuals a permanently brown colour. They protect the skin against the harmful effects of the sun's ultraviolet rays; in Caucasians they expand in response to ultraviolet light and tan the skin.

On exposure to sunlight the skin also manufactures vitamin D. And teeth, mammary glands, nails, whiskers and claws are all modified forms of skin. It is truly a multi-purpose organ.

Care of the skin

Because the skin forms the outer covering of the body it comes into contact with a wide variety of chemicals and micro-organisms. Some of these cause damage to the skin, especially chemicals manufactured by man for specific purposes such as cleaning or dyeing.

Detergents are useful because they disperse fats and oils. But sebum is an oily fluid and so it, too, can be affected by these chemicals. If the skin is regularly exposed to detergent it will become very dry, lose its suppleness, crack, and become more prone to bacterial or fungal attack; as a preventive measure one should wear protective clothing (e.g. rubber gloves) or apply a barrier cream to the skin when using detergents.

In some people a chemical coming in contact with the skin causes an allergic reaction (see page 182). This involves reddening and swelling of the skin together with itching or pain. A number of **cosmetic preparations** can irritate the skin in this way. For most people, skin allergies will arise only if the preparations are left on too long or are used very frequently over a long period.

Face powders can also be a problem; they can clog skin pores which then act as centres for infection.

So make-up should be regularly removed, the skin thoroughly cleaned and, if this process dries the skin out, a moisturising cream applied.

Similarly, hair should be carefully looked after. It should be washed regularly to prevent the growth of micro-organisms such as fungi that can cause dandruff. It is also advisable not to borrow combs or towels from other people for these can spread infections. As with the skin surface, chemicals applied to the hair, such as dyes, can be harmful. In particular, oxidative dyes, for instance peroxide, will damage hair if applied too often or left on too long.

Finger and toe nails, which are effectively compressed hair, should be kept clean and on the short side. Scratching the skin with dirty nails can lead to the entry of bacteria should the nails dig in. Finally, teeth, which also arise from the skin, should also be brushed regularly. Not doing so will lead to tooth decay (dental caries), gum disease and bad breath. Both tooth decay and gum disease can be very painful and lead to loss of teeth and serious mouth infections such as abscesses.

Key terms
Deamination Removal of amine group ($-NH_2$) from an amino acid.
Diuretic Drug that causes the production of copious dilute urine.
Egestion Defaecation; removal of undigested food from the gut.
Excretion The process of removing metabolic waste products from the body.
Homoiothermic Maintaining a constant internal temperature independent of the external temperature.
Metabolism All the chemical processes going on inside the body.
Micturition The removal of urine from the bladder.
Nephron The functional unit of the kidneys that filters the blood.
Osmoregulation Process by which the concentration of the body's fluids is maintained at a constant level.
Poikilothermic Having a body temperature that is dependent on the temperature of the surroundings.
Secretion Production of useful substances by glands, e.g. hormones and enzymes.

Selective reabsorption Retrieval of useful substances from the kidney nephrons.
Ultrafiltration In the kidneys, filtering of the blood under pressure in the glomeruli.
Vasoconstriction Contraction of muscle within the walls of a blood vessel that reduces the flow of blood.
Vasodilation Relaxation of muscle within the walls of a blood vessel that increases the flow of blood.

Chapter 6
Sensitivity and Co-ordination

Living organisms are able to detect changes in their surroundings. This ability is known as sensitivity. The human body is sensitive to a wide range of changes, or **stimuli** (singular stimulus), because it possesses millions of specialized sensory cells called **receptors.** Without receptors an organism would not be aware of its surroundings, and would not survive. It could not locate food or water, sense danger or find a mate.

Receptors are normally located on the surface of the body, for example in the eye or on the skin. Some however lie deep within the body, such as the receptors that monitor the level of blood sugar and those that monitor the body's internal temperature (see page 99).

But sensitivity alone is not enough. The body must also be able to react to what it senses. This is achieved by structures known as **effectors,** which can bring about a change in the body. Examples are the muscles and endocrine glands.

Receptors and effectors are linked together and their action is co-ordinated. This achieved by the **nervous system,** which provides the body with a highly complex and rapid communication system.

The nervous system is divided into two sections; **the central nervous system** (CNS) and the **peripheral nervous system** (PNS) (see Figure 39). The CNS consists of the brain and the spinal cord, each of which are concentrated masses of nerve cells. The PNS is a series of nerves that relay information from the receptors to the CNS and from the CNS to the effectors. The function of the CNS is to co-ordinate the information it receives from the PNS.

Complementing the activity of the nervous system are a number of slower-acting but longer-lasting chemical communicators known as **hormones,** which travel in the blood (see page 114).

Nerve cells
The nervous system is made up of many millions of nerve cells, or

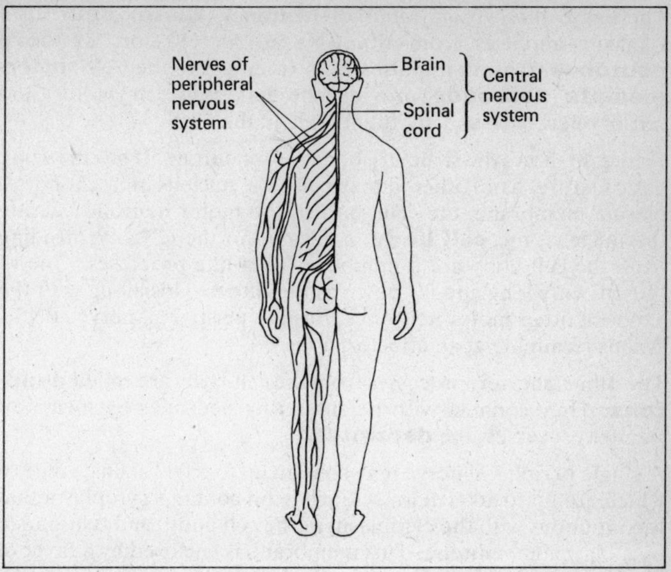

Figure 39. The nervous system

neurones. Neurones are mostly arranged in bundles joined by connective tissue to form **nerves.** Their function is to transmit messages or **impulses,** throughout the body.

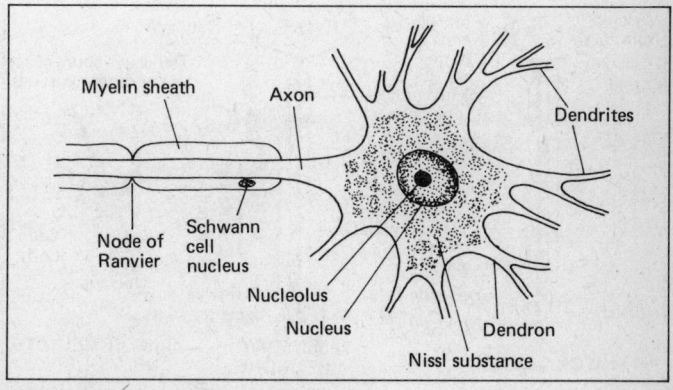

Figure 40. Structure of a neurone

There are three basic kinds of neurones. **Motor neurones** transmit impulses from the CNS to an effector; **sensory neurones** transmit impulses from receptors to the CNS; **intermediate neurones** provide the link between motor and sensory neurones and are found only in the CNS.

Figure 40 shows the structure of a motor neurone. It has the same basic features as all other human cells, i.e. nucleus, mitochondria, plasma membrane, etc. The part of the motor neurone bearing the nucleus, the **cell body,** is located in the CNS. Protruding from the cell body are a number of arm-like processes. One of these is very long and is known as the **axon.** This along with the axons of other motor neurones, forms a peripheral nerve (PNS). Axons terminate at an effector.

The other, shorter, processes from the cell-body are called **dendrons.** They connect with neighbouring neurones by means of slender processes, the **dendrites.**

A single peripheral nerve may contain up to 5,000 axons, some of which are up to 100 cm long. Each axon contains cytoplasm that is continuous with the cytoplasm of the cell body, and is bounded by a plasma membrane. This membrane is enclosed by a layer of fat, the so-called **myelin sheath.** In close connection with the axon is a **Schwann cell** the plasma of which surrounds the myelin sheath and is called the **neurilemma.** At approximately 1 mm intervals along the axon the myelin sheath is interrputed by indentations, the **nodes of Ranvier.** The sheath protects and insulates the axon and the nodes help to speed up the transmission of nerve impulses along it.

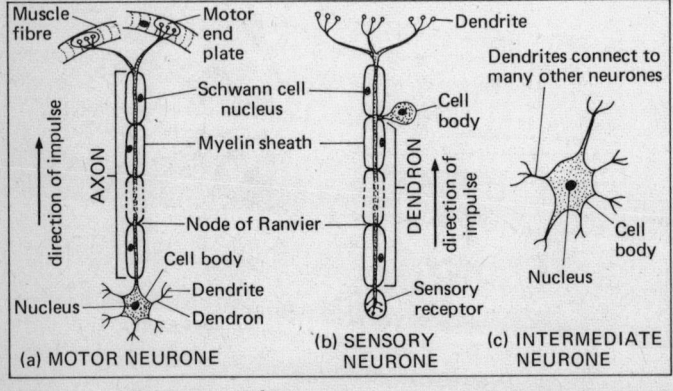

Figure 41. Three types of neurone

Figure 41 shows the structural differences between motor neurones and sensory and intermediate neurones.

The nerve impulse. In many ways the passage of an impulse along a neurone is similar to the passage of an electrical current along a wire. In fact impulses can be recorded and measured using equipment that is sensitive to small electrical currents. This has been carried out most effectively in the squid, an organism with a series of very long axons. It has been found that when the axon is resting, i.e. not transmitting an impulse, the inside of it is negatively charged and the outside is positively charged. The axon is said to be polarized, that is capable of maintaining a different electrical charge on either side of its cell membrane. When an impulse passes along the axon its polarity is momentarily reversed, i.e. the inside becomes positively charged and the outside negatively charged. A **wave of depolarization** is set up that travels quickly along the axon. This constitutes the passage of a nerve impulse. The impulse travels from node to node along the axon until it reaches the dendrites. In some axons nerve impulses can travel as fast as 100 m per second, but the speed varies from neurone to neurone.

After the passage of an impulse, there is a very short period during which another impulse cannot be transmitted. This is known as the **refractory period,** and marks the instant when the polarity of the axon is being re-established.

The synapse. The point where each dendrite of one neurone connects with the cell body of another constitutes a tiny gap or synapse. Synapses are the sites where nerve impulses pass from neurone to neurone. Passage of the impulse takes place in one direction only and is effected by means of a chemical substance, a neurotransmitter, called **acetylcholine.** Acetylcholine is produced by tiny vesicles at the tip of a dendrite. When an impulse reaches the vesicles they release the neurotransmitter which has the effect of depolarizing the neighbouring neurone and so setting off an impulse in this cell. Acetylcholine therefore acts as the bridge between two neurones.

The central nervous system (CNS)
In the embryo the CNS forms from a long tube of nervous tissue. The front (anterior) end swells up to form the developing **brain,** which later becomes differentiated into a number of distinct regions each with a specific function. The remainder of the tube becomes the **spinal cord** (see Figure 42).

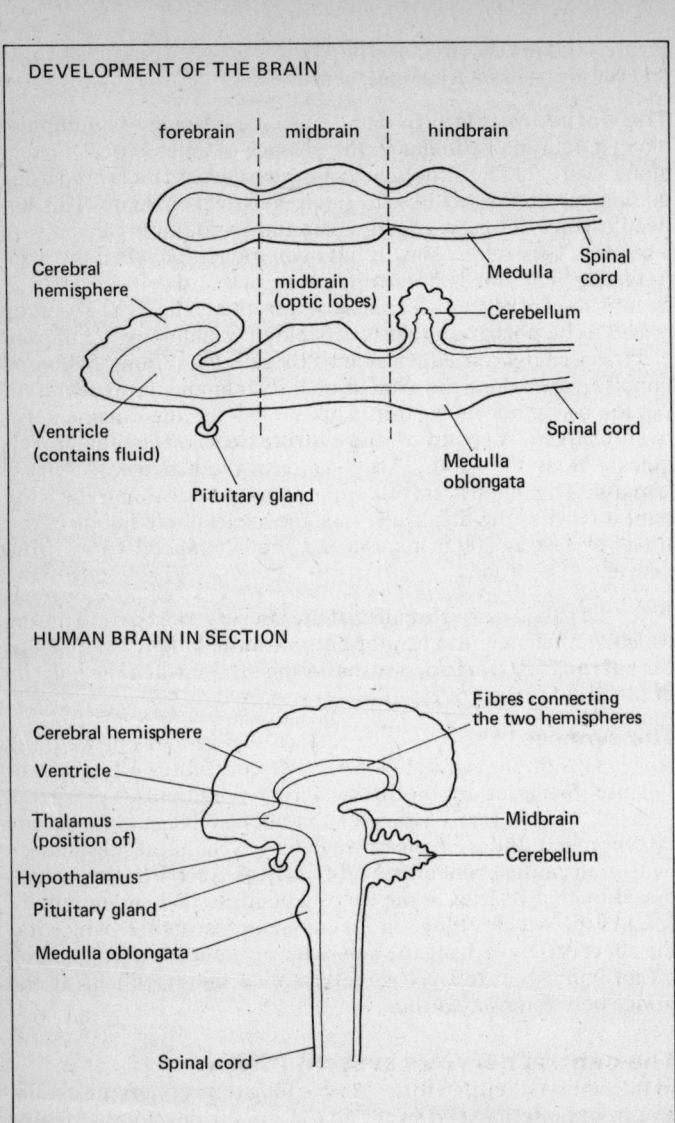

Figure 42. Development and structure of human CNS

In man the forebrain expands considerably forming two large **cerebral hemispheres** that become folded back over the rest of the brain.

The following parts of the brain can be recognized:

1. Medulla oblongata (Medulla). Contains the centres that control ventilation, blood circulation, swallowing, salivation and vomitting – all involuntary actions of the body.

2. Cerebellum. A greatly folded area of the hindbrain which controls the balance, posture and movement of the body. Man's manual dexterity is due to the highly efficient functioning of this part of the brain.

3. Midbrain. This small area contains a number of important centres. Eye movement is controlled here as is muscle tone. The anterior end contains the **thalamus,** which in relation to its size controls more actions than any other part of the brain.

Also within the midbrain is the **hypothalamus** which contains centres that control sleep and wakefulness, appetite and thirst, body temperature and osmoregulation. Beneath the hypothalamus lies the **pituitary gland,** although this does not originate from the brain. The pituitary gland secretes a large number of different hormones (see page 118).

4. Cerebral cortex. Man's forebrain is enormously enlarged in the shape of two cerebral hemispheres. By far the largest part of this area is the cerebral cortex, which is the centre of man's intellegence, memory, imagination, thought and judgement.

It is possible to record the electrical activity of the cerebral cortex by means of a machine called an **electroencephalograph.** Electrodes taped to the head pick up the activity as a wave pattern which can be amplified and recorded as an electroencephalogram (EEG). Three sorts of activity, or waves, can be distinguished. alpha, beta and delta waves. During sleep the delta waves predominate. Alpha waves are closely associated with seeing. Beta waves occur all the time. These brain waves tell us very little about the way the brain works, but they are useful in the study of people with abnormal brain functioning, such as epileptics and the mentally ill.

The tissue of the brain is among the most delicate in the body. Any damage to it is usually irreversible. Therefore it is essential that the brain is adequately nourished and protected. Nourishment is obtained through a system of three membranes, the **meninges,** which totally surround and enclose the brain and are supplied with blood vessels. Should the meninges become infected by microorganisms (meningitis) death can readily follow. Protection is afforded by the skeleton, the cranium protecting the brain and the vertebral column protecting the spinal cord.

The spinal cord

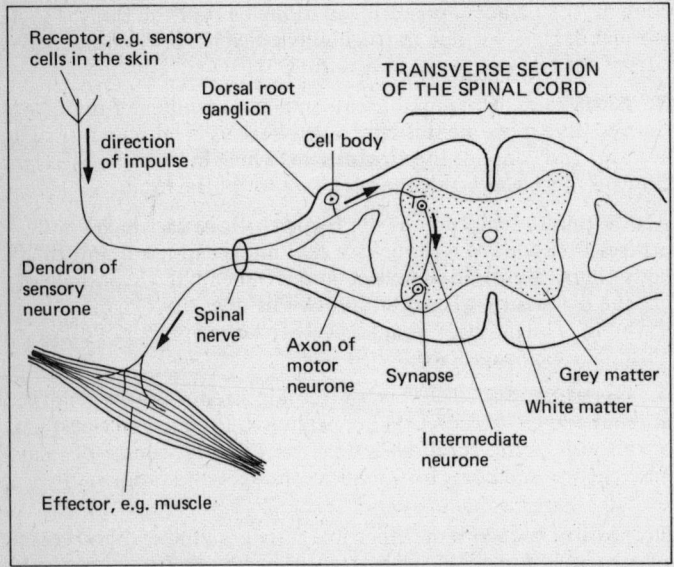

Figure 43. Reflex arc

A section through the spinal cord shows it to consist of three distinct areas. Running through the centre is the spinal canal which is continuous with the brain and is filled with cerebrospinal fluid. This contains glucose, salts, enzymes and white blood cells and nourishes the tissues of the cord. Around the spinal canal is the grey matter, which in section is H-shaped. It is made up of the bulk of the spinal cord neurones' cell bodies. The outer, white area (white matter) consists of nerve axons and dendrons. It owes its colour to the whiteness of the myelin sheaths, surrounding the axons.

The spinal cord acts as a sort of telephone exchange, linking and co-ordinating the nervous activity of the body. It relays activity between the effectors and receptors and to the brain through neurones that run up and down its length.

Reflex actions. The action of the spinal cord can be demonstrated by considering what happens when a person accidently treads on a nail.

In Figure 43, treading on a nail (an external stimulus) triggers a rapid movement of the foot away from the nail. The nervous pathway between the receptor (pressure receptors in the foot) and the effector (leg muscles) is known as a **reflex arc.** It involves a sensory, an intermediate and a motor neurone, the three connecting together in the spinal cord. The removal of the leg is an example of a *reflex* action because it does not involve any conscious thought. It happens automatically. Other reflex actions are sneezing, coughing and blinking.

In the example of a reflex arc, above, the person will also feel pain because neurones in the spinal cord send impulses to the pain centres in the brain. This makes the person aware that he has stood on a nail. But not all reflexes involve the brain; they are known as spinal reflexes. The automatic raising of the leg following a light tap below the knee (the knee-jerk) is a spinal reflex.

A conditioned reflex is an involuntary response not to the normal stimulus but to one which has been learned to be associated with it. It is acquired through imprinting on the mind that some action has a particular consequence. For example, fainting, a protective reflex action which causes a person to become unconscious and fall down so that flow of blood to the brain is restored, may be initiated by the sight of blood from a cut or wound. The person has learned that by fainting he will no longer see the blood.

The peripheral nervous system (PNS)

The peripheral nerves are divided into two groups: **the spinal nerves,** which are connected to the spinal cord, and the **cranial nerves,** which are connected to the brain. The cranial nerves are chiefly concerned with connecting effectors and receptors in the region of the head. Spinal nerves serve receptors and effectors in the rest of the body.

Spinal nerves leave the spinal cord at regular intervals along its length, each pair being connected to the cord by a dorsal and ventral root (see Figure 43). The arrangement of the cranial nerves is more irregular. This is because of the irregular shape of the brain. Man has 12 pairs of cranial nerves, which with the exception of the **vagus nerve** are confined to the head and neck. The vagus nerve is unusual as it literally wanders all over the body, passing down the neck and giving off branches to various organs including the heart and gut. It forms part of a system known as the **autonomic nervous system.**

The autonomic nervous system. This part of the PNS controls the body's involuntary activities, such as heart beat, breathing, sweating, movements of the gut and the flow of blood in the circulation. It is divided into two parts, the **sympathetic** and the **parasympathetic** systems. Both systems serve organs over which the body has little or no conscious control. The main functional difference between the two systems concerns the chemical transmitter that is released at the synapse of autonomic nerves. Parasympathetic fibres release **acetylcholine.** Sympathetic fibres release a substance called **noradrenaline,** which is very similar in structure and action to adrenaline (see page 119). The effects of the two systems generally oppose each other. In other words, if the parasympathetic fibre relaxes a certain muscle, the sympathetic fibre will cause it to contract. The general operation of the autonomic nervous system is summarized in Figure 44.

Among the many involuntary activities controlled by the autonomic nervous system are **defaecation** and **micturition** (opening of the bladder). This is surprising because most people are able to control both of these essential activities. However, if you observe the behaviour of a young baby, you will realize that they are not naturally controllable at birth. This illustrates that the various neurone connections in the nervous system are amenable to a person's willpower. You learn to control your anal sphincter by a long process of training. Other involuntary activities such as heart-rate and gut **peristalsis** can also be brought under a persons's own will after a long period of mental training. **Yoga** is one activity which can teach you to control your autonomic nervous activity.

Hormonal communication: the endocrine system

Hormones and the glands that produce them together comprise the endocrine system, which complements the nervous system as

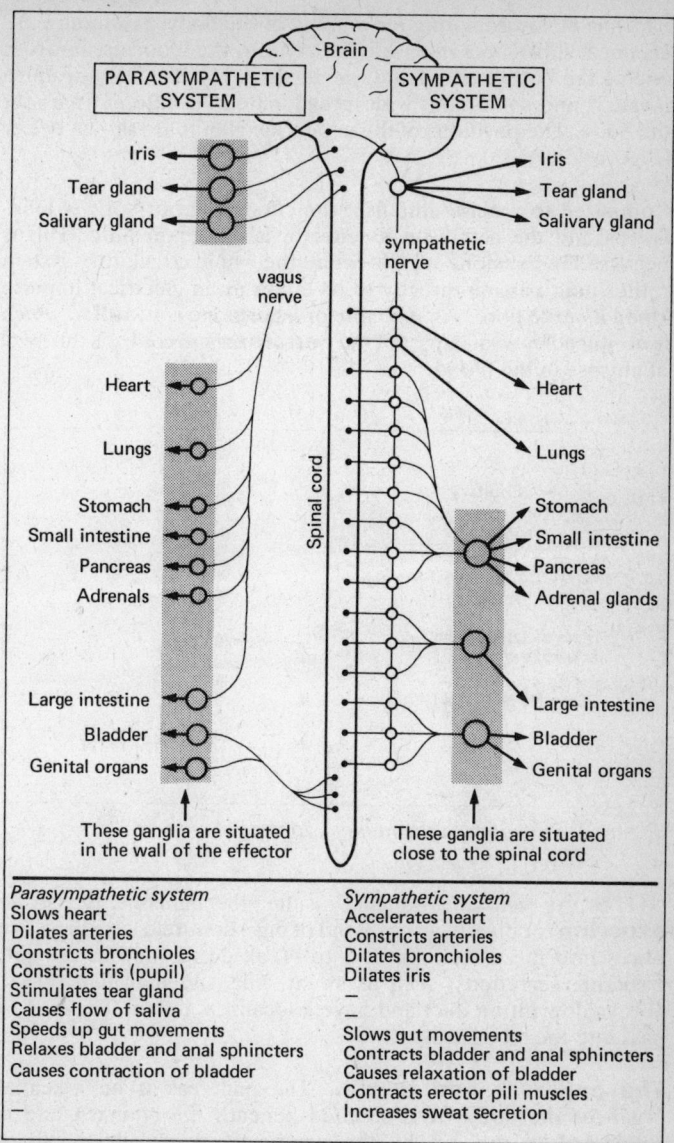

Figure 44. Summary of the action of the autonomic nervous system

a means of co-ordinating the activity of the body. Hormones are chemical substances released directly into the bloodstream from one of the body's endocrine, or ductless, glands. They produce a varied and sometimes widespread action on different parts of the body. The positions of the endocrine glands are shown below (Figure 45).

Compared to a nerve impulse, the effect of a hormone is long-lasting and the action on its effector is less immediate. This is because the hormone travels round the whole circulatory system rather than passing directly to its target as an electrical impulse along a nerve fibre. An example of a hormone is **insulin,** which is produced by a small part of the **pancreas** and controls the level of glucose in the blood.

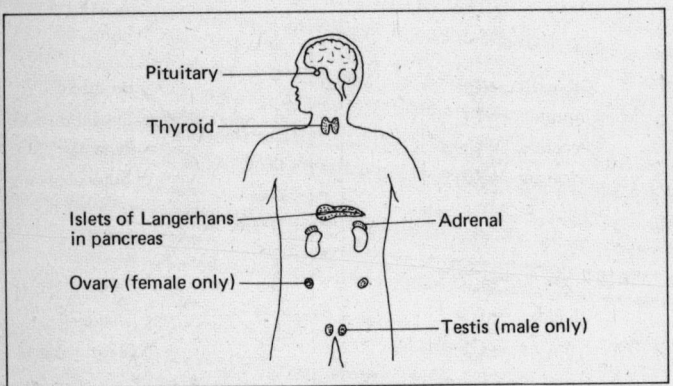

Figure 45. Position of the main endocrine glands.

NOTE: As well as producing insulin, the pancreas is also an **exocrine** (or duct-bearing) gland of digestion, releasing digestive juices into the pancreatic duct to break down food in the gut. Exocrine secretions, such as sweat, bile and digestive juices, always flow into a duct and have a localized action. They never pass into the bloodstream.

The pancreas and insulin. The pancreas is an essential organ of the body. It is situated beneath the stomach and is connected to the gut by the pancreatic duct, which carries digestive enzymes into the duodenum (see page 54).

The pancreas also secretes two hormones, **insulin** and **glucagon,** which pass directly into the bloodstream. Of the two, insulin is by far the most important. It is produced by less than 1 per cent of the cells of the pancreas, these being gathered together into tiny clumps, the **islets of Langerhans** (so-called because they were first observed by Paul Langerhans in 1869). The islets in fact contain two distinct types of cells, α and β. The β cells produce the insulin, the α cells produce glucagon.

Insulin is a complex protein consisting of two long chains of amino acids that are linked by sulphur atoms to form a stable molecule. It is produced in small quantities every day and travels to all parts of the body in the blood. The function of insulin is to assist in the general metabolism of glucose. In the liver, insulin accelerates the convertion of glucose into glycogen and throughout the body it increases the uptake of glucose by the tissues from the blood and tissue fluid.

Release of insulin is stimulated by a rise in blood glucose as occurs, for example, after a meal. Release of glucagon is stimulated by a drop in blood glucose, for instance during starvation. Glucagon thus has an opposite effect to insulin. It stimulates the liver to convert glycogen into glucose. Yet unlike insulin, glucagon is not essential to life.

Without insulin, glucose tends to build up in the blood while at the same time the tissues become starved of glucose. This can have disastrous consequences as glucose is an essential **metabolite** for cell respiration. Without it, proteins and fats are metabolized instead. The metabolism of fats increases the acidity of the blood, which can produce loss of consciousness, coma and eventually death.

The excess blood glucose is excreted by the kidneys into the urine, taking with it large amounts of the body's water. People with insufficient insulin are therefore always thirsty. They may die unless treated. Their condition is known as **diabetes mellitus,** or sugar diabetes. This is a very common disease throughout the world. In Britain alone there are more than 600,000 people diagnosed as diabetic.

Diabetes can be mild or severe. Mild diabetics produce some insulin, but not enough, so have to eat a carefully controlled diet. They may also take drugs to stimulate the pancreas to produce more insulin. Severe diabetics produce no insulin at all, so the hormone must be administered to them every day. This is usually carried out by injection with a syringe. With practice, the person learns how to do this themselves. In Britain in 1976, a special pump was developed which is worn around the waist and automatically injects the wearer with regular amounts of insulin. It acts as a sort of external artificial pancreas and controls blood glucose more effectively than syringe injections. In the USA, insulin pumps have been successfully implanted into the body so that in the future it may be possible for a diabetic to lead an almost normal life. The initial cause of diabetes, though, is still unknown.

Other endocrine organs and their hormones. The **pituitary gland** has already been mentioned. It functions as a sort of 'master gland' because many of its hormones regulate the secretion of hormones from other endocrine glands. **ADH** (antidiuretic hormone or vasopressin) controls the reabsorption of water in kidney tubules (see page 95). **FSH** (follicle-stimulating hormone) and **LH** (luteinizing hormone) play a crucial part in functioning of the male and female reproductive systems (see page 158), and **TSH** (thyroid-stimulating or thyrotrophic hormone) regulates activity of the **thyroid gland. Prolactin** stimulates breast milk production after birth and stimulates the ovary to produce the hormone progesterone (see page 159).

Of those pituitary hormones that do not control other endocrine glands are **oxytocin,** which induces contraction of the womb during labour and also stimulates the breasts to secrete milk, and **growth hormone,** which influences the growth of bone and other tissues, If the pituitary gland is underactive then growth is retarded and dwarfism results.

The thyroid gland lies in the neck on either side of the larynx (voice box). It secretes the iodine-containing hormone **thyroxine,** which controls the body's metabolic rate. Under-secretion of thyroxine during growth and development (hypothyroidism) arrests physical and mental development, a condition known as **cretinism.** This can be treated successfully with injections of thyroxine. In adults, under-secretion of thyroxine is less serious since by now growth is complete. This condition is known as **myxoedema.** Its symptoms are a decreased metabolic rate, an increase in subcutaneous fat and general sluggishness. Overproduction of thyroxine leads to **exophthalmic goitre** (hyperthyroidism), a condition marked by a characteristic swelling of the neck and protrusion of the eyeballs. Other symptoms include a fast metabolic rate, loss of weight and accelerated heart rate.

The adrenal glands lie immediately above the kidneys. Each gland consists of two regions: an outer cortex and an inner medulla. The medulla secretes the hormone **adrenaline** at times of urgency or stress. It increases the rate of breathing, the heart rate and the metabolic rate. Its action is very shortlived, but widespread. Traditionally it is called the 'flight, fright or fight hormone'. It is the hormone that makes your heart race when you are excited, or the muscles tense when faced with danger. The medulla also secretes the hormone **noradrenaline.** Its effects are very similar to **adrenaline.** Noradrenaline has been mentioned earlier as a synaptic chemical transmitter substance in the sympathetic nervous system. In fact, the adrenal medulla is stimulated by nerve impulses from the sympathetic system.

The adrenal cortex is stimulated by pituitary hormones, in particular adrenocorticotrophic hormone (ACTH), to produce three kinds of **corticosteroid** hormones, one of which affects carbohydrate metabolism and another the development of the reproductive organs.

The reproductive organs (the ovaries and testes) also produce hormones. The ovary secretes oestrogen and progesterone (see page 153). The testes secrete testosterone and a little progesterone (see page 158).

The stomach also produces a hormone, **gastrin.** This is secreted when food begins to enter the stomach and stimulates the flow of gastric juice (see page 52).

Hormones and nerve impulses compared. The endocrine and nervous systems both provide a means of communication within the body. Both transmit messages which are triggered by a stimulus and both evoke a response. The main difference between the two systems is in the nature of the message. These differences are summarized below.

Nervous system	Endocrine system
1. Nerve action is rapid and short-lived.	1. Hormones work slowly, tending to regulate continuous or long-term processes such as growth and metabolism.

2. Nerve responses tend to be very localized. A muscle moves, or an eye blinks.	2. The action of a hormone is sometimes widespread (as in the case of adrenaline), but on the whole each hormone has a particular target organ (c.f. pituitary ADH).
3. Message is carried as an electrical charge along a nerve fibre.	3. Message is a chemical substance carried in the bloodstream.
4. Nerve impulses are all-or-nothing in their action. In other words, either an impulse travels or it doesn't.	4. Hormones can be secreted in varying amounts. Their effects are directly related to the quantities that are produced. An excess or deficit of hormone produces a corresponding characteristic disease.

Reception of stimuli: sensitivity

Receptors consist of sensory cells that have the task of converting stimuli into electrical nerve impulses. The sensory cells may be found singly, scattered more or less evenly throughout the body, or concentrated to form a sense organ.

Receptors can be classified according to the type of stimulus they respond to.

1. Chemoreceptors. Stimulated by the presence of chemicals, for example those detecting smell in the nose and taste on the tongue.
2. Mechanoreceptors. Respond to mechanical deformation, for instance those stimulated by touch, pressure, stretch, sounds (pressure waves) and movements of the body (balance).
3. Photoreceptors: Stimulated by light, for example receptors in the eyes.
4. Thermoreceptors. Respond to changes in temperature, such as the internal receptors that monitor the temperature of the blood.

Receptors are further classified according to whether they respond to external or internal stimuli. These are known as **exteroceptors** and **interoceptors** respectively.

Sensory cells. The structure and mode of action of these cells varies according to the nature of the stimulus they respond to. Some are neurones only slightly modified in the shape of a sensory terminal at one end. Others consist of specialized neurones that have one or more slender processes adapted to register changes in a particular environmental factor. This stimulus is then converted into a nerve impulse.

In the light-sensitive retina in the eye, there are two types of sensory cells, the **rods** and **cones.** Both are stimulated by visible light, but the rods are more sensitive and work best at low levels of light, i.e. at night; the cones are concerned primarily with colour vision.

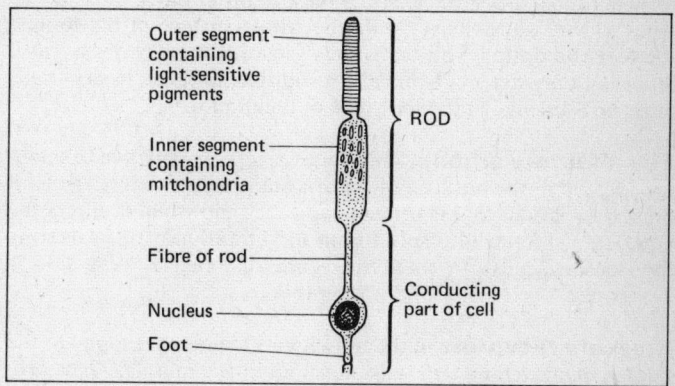

Figure 46. A human receptor cell; a rod

The structure of a rod is shown above (Figure 46). Its outer segment contains a coloured pigment, **rhodopsin,** which is broken down in the presence of light. This 'photochemical' reaction evokes a depolarization of the cell membrane and generates an electrical impulse at the foot of the cell. Rhodopsin is then immediately resynthesized using energy from the many mitochondria inside the rod's inner segment. Most people will have experienced the effects of rhodopsin breakdown and resynthesis. If you enter a darkened cinema from the brightness of the cinema foyer, at first you can see nothing. But gradually you stop stumbling about as you begin to see more and more of your surroundings. This is because your rhodopsin, broken down by the brightness of the light in the foyer, is gradually being regenerated. This process is known as **dark adaptation.**

Rhodopsin is a protein composed of a light-sensitive substance called **retinene,** which is derived from carotene (vitamin A). Hence the importance of vitamin A in the diet.

In the skin there are a number of different types of receptors. Free nerve endings are sensitive to pain, **Meissner's corpuscles** under the epidermis are sensitive to touch, while the deeper **Pacinian corpuscles** are sensitive to pressure. In addition, temperature receptors are sensitive to heat and cold (see page 99).

On the tongue are **taste cells.** These are found in groups called taste buds, and there are four kinds, each sensitive to a particular type of substance. Taste buds at the tip of the tongue are sensitive to sweet substances, those at the back to bitter ones, those at the sides to sour substances, while the whole surface of the tongue seems to be dotted with taste buds sensitive to salty ones. Taste buds can only detect chemicals in solution so those in dry foods must first dissolve in the moisture of the mouth.

The olfactory organs, which bear receptors sensitive to smell, are located in the nasal cavity. For a substance to have a smell it must be volatile, that is it must vaporize, and when it enters the nose it must enter into solution in the film of moisture covering the sensory cells. In man the sense of smell is very poorly developed.

Pressure receptors in the muscles and tendons respond to the degree of stretching and tension set up in the muscles. This provides information about the position of the limbs and the muscles. This is essential for co-ordinated movement.

Sense organs: the eye and the ear

The location and general structure of these two highly specialized organs are shown in Figure 47. The position of the ears and the shape of the external ear, or pinna, allow an almost complete collection of sound from every direction. The frontal position of the eyes allows them to function in unison. This fact has been important in the evolution of the primates, man in particular, because together the two eyes provide stereoscopic, or 3-D, vision. The result is that individual objects are clearly distinguishable and can be seen in relation to the position of others. Each eye sees a slightly different image, but together an impression of distance is achieved. This ability can be appreciated by closing one eye and trying to work out how far one object is behind another.

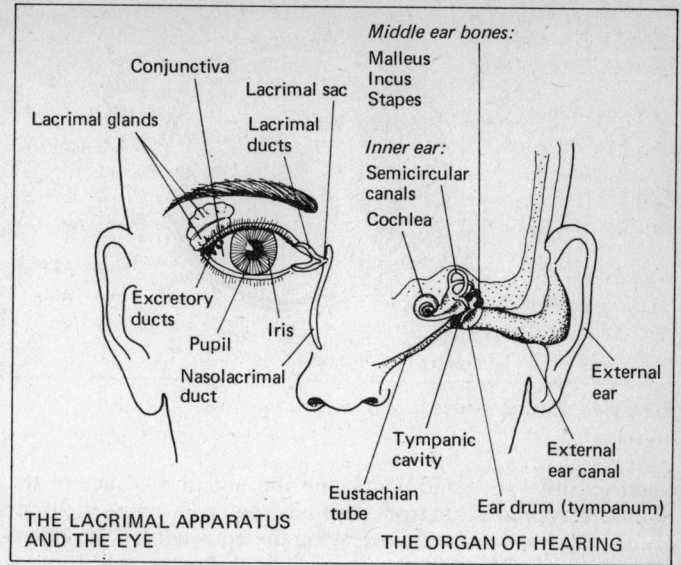

Figure 47. Location and structure of the human eye and ear

Structure of the eye. Each eye is held in a bony socket of the skull by six eye muscles. The front of the eye is protected partly by the eyelids and eyelashes. Under the eyelids are tear glands which secrete a fluid that washes away dust particles and destroys bacteria. The **conjunctiva** is a thin layer of tissue lining the eyelids and covering the front of the eye. Here, beneath the conjunctiva is a thick, transparent layer, the **cornea**. The sclerotic coat is a continuation of the cornea. It forms a thick protective layer round the eyeball. The anterior cavity of the eye is filled with water-like **aqueous humour** and the posterior cavity with jelly-like **vitreous humour** (also known as the vitreous body). These fluids together maintain the shape of the eyeball. Behind the cornea is the **iris** in the centre of which is a space, the **pupil,** through which light passes to the retina. The iris contains blood vessels, muscles and colouring pigment. The lens is held in position by **suspensory ligaments** attached to the circular **ciliary body.** The function of the lens is to focus light onto the **retina.** The iris is continuous with the pigmented **choroid layer** which nourishes the retina. Figure 48 shows the detailed structure of the eye.

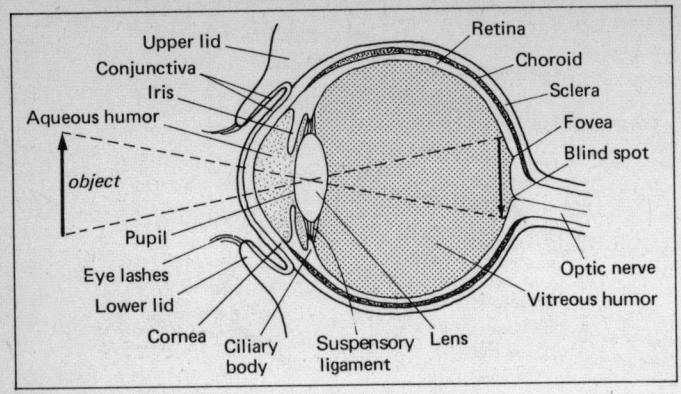

Figure 48. Detailed structure of the human eye

Light-sensitive rods and cones line the innermost layer of the retina. Nerve fibres run from their base and pass across the front of the retina and leave the eye along the optic nerve. The **blind spot** is the point where the nerve fibres leave the eye. It is so-called because it contains no sensory cells and so cannot respond to light. (You are not aware of this though because the visual fields of your two eyes overlap.)

The muscles of the iris control the diameter of the pupil and therefore how much light falls on to the lens. In bright light a ring of muscle round the margin of the iris contracts, causing the pupil to become smaller. In dim light radiating muscles contract, increasing the size of the pupil. The suspensory ligaments act to change the shape of the lens so that it focuses the light on to the retina to produce a sharp image. If you look from a close object to a far distant object, you can almost feel your suspensory ligaments working. The process of the lens changing its shape to focus on an object is called **accommodation.**

The **fovea centralis,** or yellow spot, is an area in the centre of the retina which is dense in cones. Cones detect colour but only work in bright illumination. They contain visual pigments that are not readily bleached even by bright light, unlike the rhodopsin found within the rods.

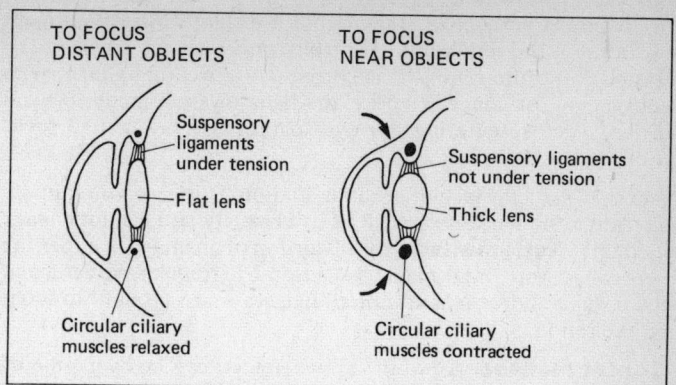

Figure 49. Accommodation of eye

Cones also provide the eye with its ability to distinguish between objects. This ability is known as **visual acuity,** or sharpness of vision. In comparison with other animals, the mammalian eye has a high visual acuity. To illustrate this, hold a book about 30 cm in front of you and look directly at it. You should be able to distinguish all the letters on the page, even the tiniest. Now, keeping your head and eyes looking in the same direction, move the book 30 cm or so to the left or right. Without directly looking at the book, try to read it. Of course you cannot; all you can see is a blurred image of something black on white. This is because the light from the book is no longer striking the retina at the fovea but at an area which is dense in rods but has very few cones. The same exercise can be tried using colours. If you do not look directly at a coloured object, you cannot perceive its colour. Test this for yourself. The explanation again is that the cones, which detect colour, are found in sufficient numbers only in the fovea. To detect colour, light must fall on to this part of the retina.

Defects of vision
Short sightedness (myopia) is usually caused by the eyeball being too long from front to back. Light from a near object is brought to focus on the retina but light from distant objects is brought to a focus in front of the retina so the object appears indistinct. It is corrected by wearing glasses with diverging, or concave, lenses; the rays of light are caused to bend outwards before entering the eye and so are focused as they should be.
Long sightedness (hypermetropia) occurs in people who cannot focus light from near objects. It is caused by a weak lens

or too short an eyeball. Light from a distant object is brought to focus on the retina but light from a near object is brought to focus behind the retina. It is corrected by wearing glasses with converging, or convex, lenses; the light rays are made to bend inwards before entering the eye so they are brought to focus properly on the retina.

Another defect is lack of accommodation. This is common in old age when the lens loses much of its elasticity and cannot change its shape. **Astigmatism,** the distortion of an image in either the vertical or horizontal plane, is caused by irregular curvature of the cornea and/or lens so not all light rays are brought to focus on the retina.

Contact lenses. All of the above defects can be overcome by wearing spectacles with lenses of differing convexity or concavity according to the extent of the problem. Contact lenses are an alternative to spectacles. They are made of glass or plastic and are fitted over the pupil of each eye. They are invariably much more troublesome to wear than spectacles, but as they are largely invisible, they may improve some peoples' appearance.

'Soft' (plastic) contact lenses are easier to wear than 'hard' (glass) ones, but are more expensive and require greater maintenance. Soft lenses will snap unless kept in water, and because they absorb water, people with dry eyes are not able to wear them. They can also absorb fumes or chemicals in the atmosphere and bring them into contact with the highly sensitive tissue of the eye. However, there is less risk of breakage and subsequent damage to the eye than with hard lenses.

The ear

The mammalian ear consists of three chambers: an **outer,** a **middle** and an **inner.** The outer and middle ear are filled with air and the inner ear mainly with a fluid called **perilymph.**

The middle ear, separated from the outer ear by the **tympanic membrane** (tympanum or eardrum), contains three tiny bones, the **ear ossicles.** They are named according to their shape: **malleus** (hammer), the **incus** (anvil) and the **stapes** (stirrup). They are in direct contact with each other and are held in place by ligaments. One of them, the malleus, is in contact with the eardrum, and another, the stapes, is in contact with the entrance to the inner ear, the oval window. The **Eustachian tube** connects the middle ear with the pharynx so that the air pressure on both sides of the ear drum is equal.

The outer ear consists of the **pinna,** or earlobe, a flexible funnel which channels sound into the (external) **auditory canal** and towards the ear drum. The lining of the canal bears secretory cells that produce a thick wax to prevent the entry of micro-organisms and other foreign particles.

Description of the structures within the inner chamber is best related to the two basic functions of the ears, namely hearing and balance.

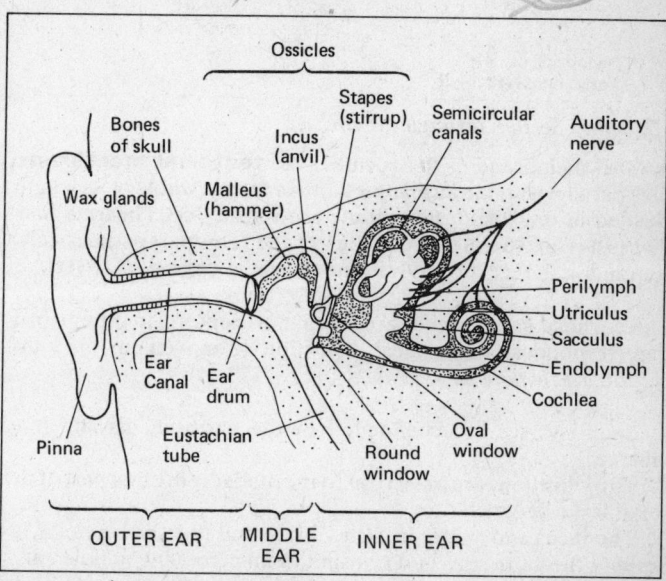

Figure 50. Structure of the ear

Hearing. In section, the **cochlea** of the inner ear is seen to contain three chambers, or canals. The upper and lower canals are filled with **perilymph.** The middle canal is filled with another fluid, **endolymph,** and is separated from the other canals by two membranes (see Figure 51), the **vestibular membrane** and the **basilar membrane.** The upper (vestibular) canal is in contact with the oval window, which is essentially a membrane. The lower (tympanic) canal is in contact with the round window, again a membrane.

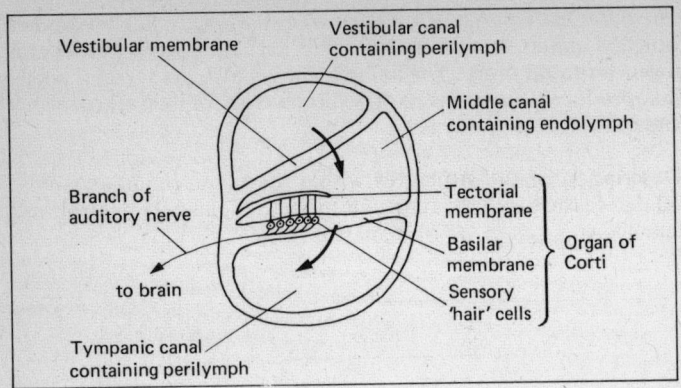

Figure 51. Section through the cochlea

A third membrane in the cochlea, the **tectorial membrane,** runs parallel with the basilar membrane for the whole of its length. Nestled in the basilar membrane are sensory cells bearing hairs that just reach to the tectorial membrane. This part of the cochlea containing the sensory cells is known as the **organ of Corti.**

The cochlear sensory cells are mechanoreceptors, in other words they respond to mechanical distortion or movement. The distortion is achieved as follows:

1. Sound waves are channelled to the eardrum causing it to vibrate.
2. The vibrations are passed on to the malleus, the first part of the small lever of ossicles.
3. The incus and stapes in turn vibrate and together the ossicles amplify the vibrations and transmit them across the middle ear.
4. Vibration of the stapes causes the oval window to move backwards and forwards, which sets up movements (fluid waves) in the perilymph contained within the vestibular canal.
5. This in turn causes movement of the vestibular membrane such that fluid waves are set up in the endolymph in the middle canal.
6. The basilar membrane is thus made to vibrate, thereby displacing fluid in the tympanic canal.
7. These fluid movements stretch the membrane of the round window.

It is movement of the basilar membrane that causes the hairs of the sensory cells of the organ of Corti to become pushed against the tectorial membrane and thus distorted. This mechanical stimulus results in the sensory cells generating nervous impulses, which pass along the auditory nerve to the brain.

The precise mechanism of stimulation of the sensory cells is not known. Research is hampered by the fact that this part of the ear is very small and is deeply encased by bone.

Balance. The three semicircular canals of the inner ear are filled with endolymph and contain sensory structures that detect movement of the head. At one end of each canal is a swelling, the **ampulla,** which contains a receptor. This consists of a group of sensory cells whose hairs are embedded in a gelatinous mass, the **cupula.** The semicircular canals are arranged in three different planes so that movement in any direction will be detected. A tilting, turning or sudden raising or lowering of the head causes the endolymph in the canals to move towards the ampulla and so displace the cupula; this stimulates the sensory cells and impulses are passed along the auditory nerve to the brain.

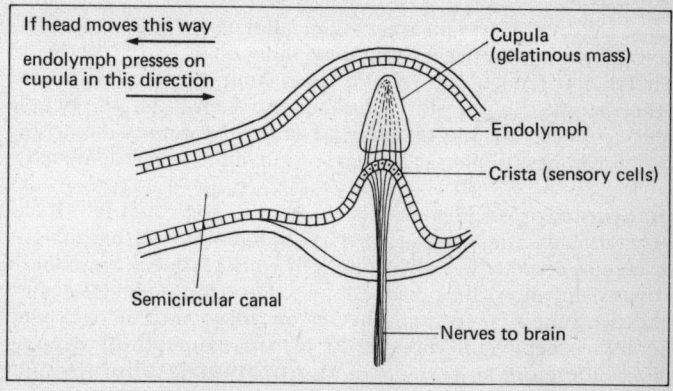

Figure 52. Section through ampulla

Finally, within the **utriculus** and **sacculus** of the inner ear are receptors consisting of sensory cells, the free ends of which are embedded in a granule of calcium carbonate, an **otolith**. Any change in posture will tend to displace each otilith and increase or decrease the pull on the hairs of the sensory cells. This stimulates the neurones and impulses are transmitted to the brain giving information about the new position of the head in relation to gravity.

Sleep

Most people spend about a third of each day asleep. Yet although sleep appears to be a very necessary activity, very little is known about its function.

During sleep, as is mentioned on page 111, the activity of the cerebral cortex changes from alpha and beta waves to delta waves. But although the cortex is active we are not conscious of it. The switch-over from wakefulness to sleep is poorly understood. It seems to have its own day-and-night mechanism but this can be readily modified, as all jet travellers know to their cost.

The stimulus for sleep appears to be reinforced by a reduction of sensory information, i.e. by comfort, quietness and darkness. But nobody knows why the brain should need a rest.

Certainly during sleep the chemical processes of the body are slowed and the flow of saliva, tears, mucus and urine are also reduced.

Most children sleep a lot longer than adults, but there is considerable variation between different individuals. Mrs Margaret Thatcher is reported to require fewer hours of sleep than most other people. Sir Winston Churchill also slept little. Old people tend to sleep less, but this may be caused by depression arising from loneliness or physical deterioration of the brain (dementia).

Insomnia, or sleeplessness, is a very common condition. It can result directly from pain, such as from indigestion, or from anxiety. A common remedy is a sleeping pill which reduces the pain or anxiety and allows the person to sleep. However, unless the cause of the insomnia is removed, the person will soon come to depend upon the sleeping pills in order to sleep. Treatment with sleeping pills is therefore an example of **symptomatic relief.** It is not treating the illness or the problem but merely the symptoms.

Drugs and the nervous system

A drug is a substance which can change the way the body works physiologically. In general, people take drugs as part of a medicine which can either help them recover from an illness, relieve symptoms such as pain, or to modify any natural process in the body.

Every medicine contains a drug, but not every drug is a medicine. **Alcohol, nicotine, caffeine** and **cannabis** are four of the most commonly taken drugs, but rarely, if ever, are they taken as medicines.

Over 20,000 different medicines are available in the UK today and more than £750 million will be spent on drug prescriptions in the UK in a year (1982). The manufacture of drugs by the pharmaceutical industry is very big business, but here we shall concentrate on one small aspect only, namely the effects on the nervous system of two small groups of drugs, the **stimulants** and **sedatives.**

Stimulants and sedatives are examples of **psychotropic** or mood-altering, drugs. Psychotropic drugs are classified according to their effects. **Hypnotic** drugs, for example, produce sleep. Sedatives calm you without sending you to sleep. In large doses most sedatives are also hypnotics. A **tranquilizer** is a mild sedative that calms you without affecting your conscious state. A stimulant raises your mood, or gives you a 'lift'.

Sedatives such as alcohol or barbiturates depress the functioning of the brain. The higher the dose, the greater the reduction in function and hence the influence on behaviour.

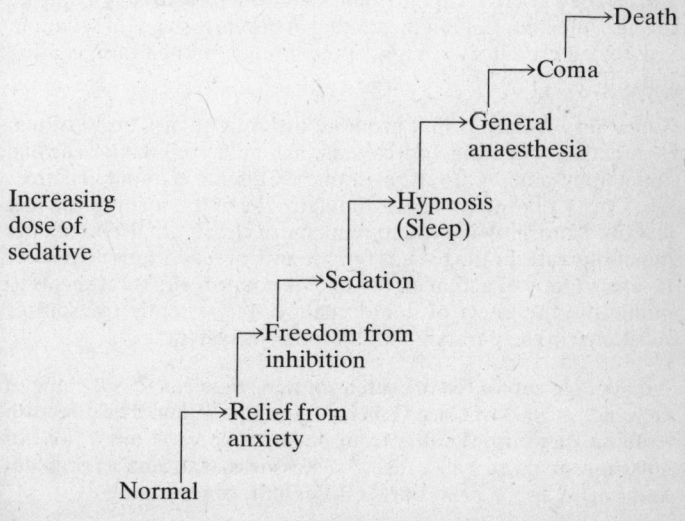

Sedatives effect the activity of the brain by slowing down the transmission of nerve impulses across synapses. (They do not affect the transmission of impulses along the axons of neurones.) The thalamus, the part of the brain that controls wakefulness, is particularly packed with synapses and this accounts for the general effects of sedatives in normal doses. At higher doses, sedatives also depress the activity of the heart, the PNS, the muscles and the metabolic rate, eventually causing death.

The taking of sedatives over a long period can induce psychological and/or physical **dependence** and **tolerance.** Dependence is experienced as an urgent need to take the drug again. In physical dependence, to stop taking the drug leads to specific symptoms **(withdrawal symptoms)** that are alleviated by further doses. Tolerance involves an adaptation by the body to the drug, such that the neurones of the brain learn to function normally only in the presence of ever-increasing amounts of the drug. Tolerance is particularly dangerous in the case of alcohol because in large doses this is cumulatively toxic to the liver, leading eventually to death.

If you have ever drunk any alcohol you will know that rather than sending you to sleep, it makes you feel excited and less inhibited. So why should alcohol be classed as a sedative? The explanation is simple. To begin with all sedatives at first depress only the **inhibitory** synapses in the brain. Later the **excitatory** synapses are also affected. But this means that in the early stages of sedation, unaffected excitatory synapses predominate in the brain, causing the 'lift'.

Stimulants are drugs that promote the activity of a body system or function. Nicotine and caffeine are mild stimulants, cocaine and amphetamines are stronger ones. Caffeine is found in coffee, tea, Coca Cola and cocoa. It stimulates the brain, making you feel less tired and allowing you to think more clearly. It also raises the metabolic rate and so lessens fatigue and increases muscle power. Its exact mode of action on the CNS is not proven, but it seems to mimic the the effect of noradrenaline, the synaptic transmitter substance in the parasympathetic nervous system.

An average cup of tea or coffee contains between 75 – 125 mg of caffeine. A glass of Coca Cola contains 35 – 55 mg. The effects of caffeine vary considerably from person to person but a dose of 1000 mg or more will cause sleeplessness, extreme excitement and ringing in the ears. Larger doses still, can be fatal.

Psychological dependence and tolerance to caffeine is commonplace. Most people cannot pass the day without their daily supply of tea or coffee (or Coca Cola).

Key terms

Accommodation Reflex action causing the focal length of the lens of each eye to alter, enabling it to focus on a near or far object.
Autonomic nervous system Part of the PNS which controls the involuntary activity of the body.
Brain Enlarged anterior region of the CNS.
Central nervous system (CNS) The brain and spinal cord.
Cerebellum Expanded region of the hindbrain responsible for posture and balance.
Cerebrum Expansion of the forebrain associated with memory, intelligence and thought.
Cochlea Part of the inner ear which contains receptors to the vibrations of sound.
Depolarization A change in state of the electrical charge across a cell membrane.
Endocrine gland A mass of cells which secrete a hormone into the bloodstream.
Exocrine gland A mass of cells which secrete substances directly into the surroundings via a duct.
Homeostasis Maintenance of a constant internal environment.
Hormone A chemical produced by an endocrine gland and secreted directly into the bloodstream which affects a specific, usually distant, part of the body.
Hypermetropia Long-sightedness; inability to focus objects close to the eye.
Insomnia Sleeplessness.
Medulla oblongata Part of the hindbrain which controls breathing movements.
Myopia Short-sightedness; inability to focus objects a long distance from the eye.
Neurone A nerve cell.
Peripheral nervous system (PNS) The spinal and cranial nerves.
Receptors Nerve cells that respond to a stimulus by generating an electrical impulse.
Reflex action Rapid automatic response to a stimulus.

Sedative A drug that will depress the nervous activity of the brain.
Sensitivity The ability to detect changes in the internal and external environment.
Spinal cord Part of the CNS protected by the vertebral column.
Stimulant A drug that will promote the activity of the brain.
Stimulus Any change in the internal or external environment.
Synapse The place where two neurones meet.

Chapter 7
The Skeleton, Locomotion and Growth

Locomotion is the movement of an organism from one place to another. It is brought about by movements of parts of the body, a process that in man involves the co-ordinated contraction of muscles against the bones of the skeleton. Locomotion enables man to obtain food, avoid predators, seek out mates and disperse to new habitats. A skeleton provides the body with a rigid supporting structure, as well as acting as an anchor for the muscles and protecting its more delicate organs.

The human skeleton is an example of an **endoskeleton.** This means that the hard supporting material, in this case bone and cartilage, is found inside the body. **Exoskeletons,** such as those belonging to insects, crabs, lobsters, etc., have the hard material on the outside of the organism. A third type of skeleton, called a **hydrostatic skeleton,** is found in organisms such as the earthworm and the jellyfish. Here fluid inside the body is maintained under pressure and is surrounded by muscle bands which contract against it. Each type of skeleton affects the way in which an organism grows (see page 144).

The human skeleton
The skeleton contains 206 separate bones. A third of them are found along the axis of the body, forming the so-called **axial skeleton.** This comprises the skull and the vertebral column. The remaining bones are suspended from the axial skeleton and together form the **appendicular skeleton.** This consists of the pectoral and pelvic girdles together with the bones of the arms and legs (see Figure 53).

The vertebral column is approximately 700 mm long in the adult and is made up of 26 bones (see Figure 54). Immediately beneath the skull are the seven neck, or **cervical,** vertebrae. The first of these, the **atlas,** and second the **axis,** carry the weight of the head and allow it to turn. Next are twelve **thoracic** vertebrae that provide attachment for the ribs. Then there are five **lumbar** vertebrae, to which are attached the strong back muscles. The five **sacral** vertebrae are fused to form the **sacrum** which is part of the pelvis. The four **caudal** vertebrae are also fused, forming the **coccyx,** the so-called 'tail'.

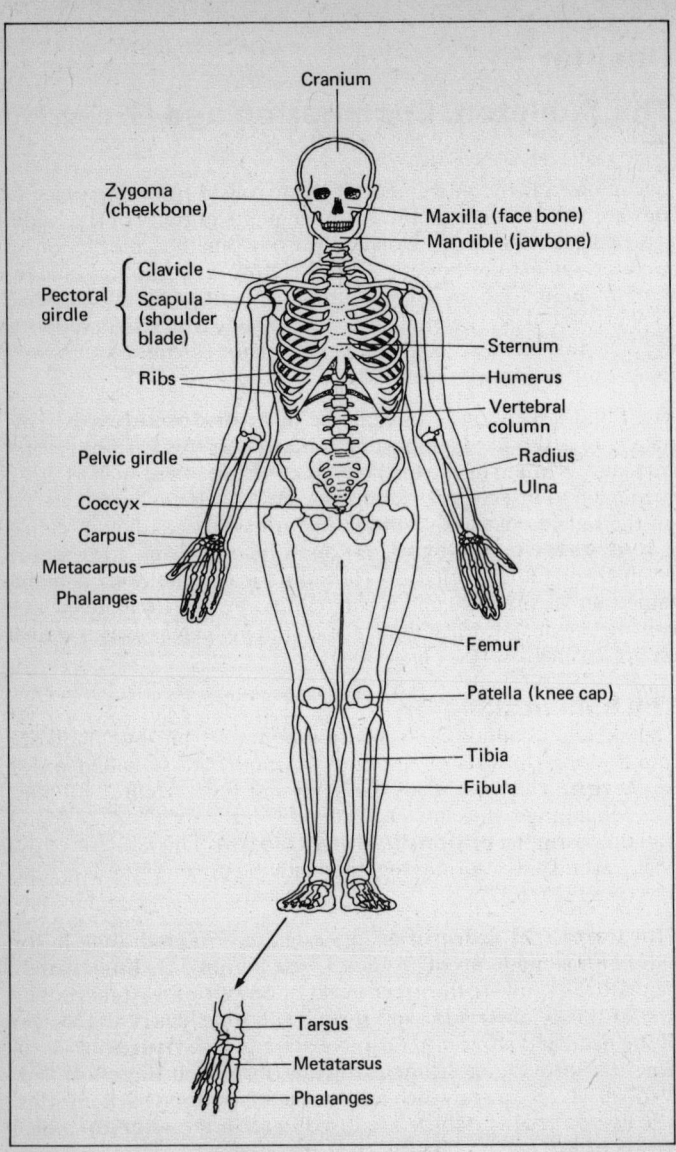

Figure 53. The human skeleton

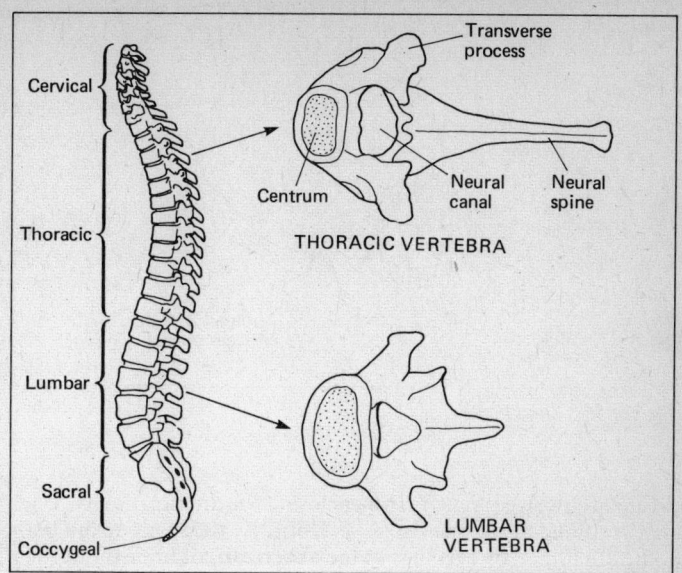

Figure 54. The vertebral column

The skull consists of 22 bones, which apart from the lower jawbone, or mandible, are fused together. It includes the **cranium,** which contains and protects the brain. In the adult the cranial bones are fused by the interlocking of fine jigsaw processes along their margins to form joints called **sutures.** These are not completely formed in the newborn, allowing the head to squeeze out of the mother's pelvis during birth and for subsequent growth.

Through several holes in the base of the skull pass nerves and blood vessels. The largest of these bears the spinal cord, the main thoroughfare for nerve impulses passing to and from the brain and the rest of the body. The spinal cord passes through and down the **neural canal** of the vertebrae. Fourteen facial bones are attached to the cranium. They support the muscles of the face, mouth and nose. The upper jaw consists of two **maxillae,** which meet at the midline below the nose. The lower jaw articulates with the cranium in front of the ears. The maxillae and mandible both carry teeth.

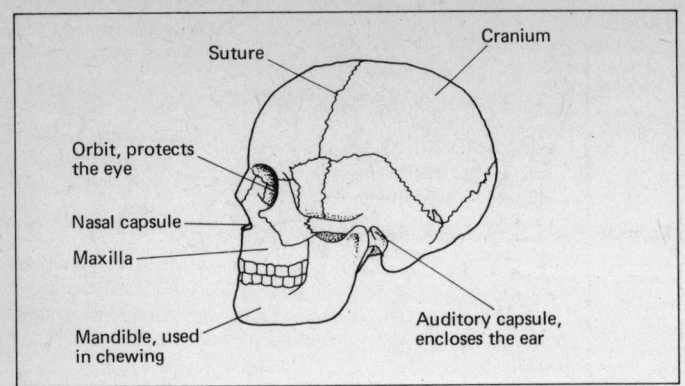

Figure 55. The skull

There are twelve pairs of **ribs** enclosing the thoracic cavity. Each pair is joined at the back to a vertebra. At the front of the body some are joined by cartilage to the **sternum,** or breastbone. The **pectoral girdles** each consist of a **clavicle** (collar bone) and **scapula** (shoulder blade) and project out from behind the rib cage so that the arms can swing freely. Each arm contains three long bones. The upper arm bone, the **humerus,** forms a movable joint with the **scapula.** At the elbow, the **radius** and **ulna** join the humerus, the radius on the thumb side. There are nine small wrist bones, the **carpals.** They join five long **metacarpals,** the bones in the palm of the hand. The finger, or digit, bones, **phalanges,** number three in each finger and two in the thumb. The thumb is able to oppose (touch) the tips of each finger and this enables a wide range of manual skills to be performed.

The **pelvic girdle,** or pelvis, is a complete ring of bone formed by the sacrum at the back and the hip bones at the sides and front. The hip bones are made up of three fused units, the ilium, ischium and pubis. The pelvis surrounds and protects the bladder and rectum, and the uterus in females. It forms a movable joint with each **femur** (thigh bone) which in turn joins the **tibia** and **fibula** (lower leg bones) at the knee. The inner of the two, the tibia, is shorter than the more delicate fibula, and both are slightly shorter than the femur. The knee joint is protected by a small cap of bone, the **patella.** The stout ankle bones, the **tarsals,** join the long **metatarsals** which give the foot its length.

The forelimbs enable objects to be lifted and manipulated. The hind (lower) limbs essentially support the body and permit locomotion.

Joints. A skeletal joint is where the ends of two or more bones meet. They are grouped according to the type of movement they allow:

1. Ball and socket joints allow movement in many planes – where the top of the head, or 'ball', of the humerus moves in the socket of the scapula, for example.
2. Hinge joints allow movement in one plane only – the elbow and knee joints.
3. Double-hinge joints allow movement in two planes – the wrist joint, for instance.
4. Gliding joints allow a sliding movement of one surface over another, as is possible between the bones in the wrist and ankle and between the vertebrae.
5. Immovable joints exist between bones that are fused. The best examples are the suture joints between the bones of the cranium.

The first three types of joint are freely movable and are known as **synovial joints** (see Figure 56).

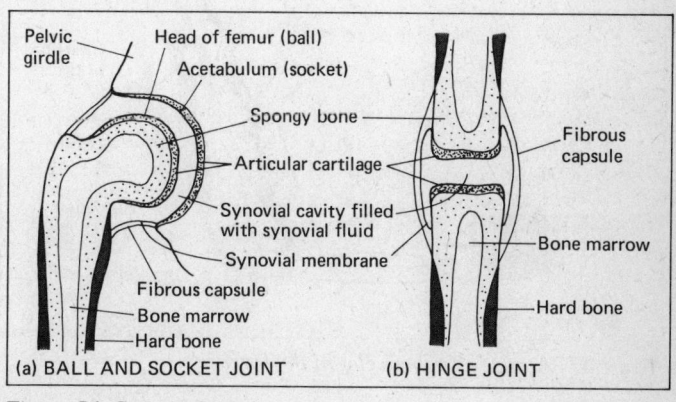

Figure 56. Synovial joints

The ends of two bones concerned are held firmly by **ligaments** that form a surrounding **capsule.** This is lined by the **synovial**

membrane which secretes **synovial fluid.** The fluid bathes the joint helping the bones to move smoothly by acting as a lubricant. Layers of cartilage stretch down and across the ends of the bones forming a buffer between them.

Movement

Movement of the skeleton is brought about by muscles attached to the bones on either side of a joint (see Figure 57). The function of the muscles is to contract and pull one bone towards or away from another. In a movable joint, the bones are connected to one another by tough connective tissue ligaments, which are elastic and suited to bearing sudden stresses. The muscles are attached to the bones by **tendons,** which are non-elastic but also capable of bearing sudden stresses. The muscles themselves, of course, have considerable elasticity. All the muscles have their origin on a bone which is immovable when the muscle contracts and their insertion on a movable bone that is situated further away from the centre of the muscle. The contraction of a muscle therefore normally pulls the point of insertion towards the origin. This mechanical movement is brought about by means of leverage; the bones act as levers through which the power of muscles is used. In Figure 57, the joint acts as the fulcrum.

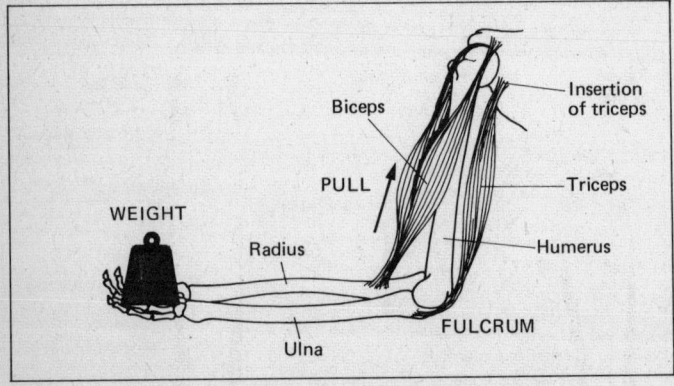

Figure 57. Antagonistic muscles of the forearm

The elbow is a synovial joint between the humerus, radius and ulna. The muscles responsible for movement about this joint are the **biceps** and the **triceps.**

The biceps lies in front of the humerus. At one end it is divided in two and attached by tendons to the shoulder. At the other end it is attached to the radius bone, a little below the elbow joint. The triceps lies behind the humerus. Tendons attach the muscle, at one end to the shoulder, and at the other end to the ulna.

Bending the arm (flexion) to lift an object involves the following action. The biceps contracts and pulls the radius upwards. The triceps relaxes as this occurs. The biceps muscle producing this bending is known as a **flexor** muscle.

Straightening the arm (extension) involves the opposite action. The triceps contracts, pulling the ulna downwards, while the biceps relaxes. The triceps muscle thus acts as an **extensor** muscle.

Flexor and extensor muscles are **antagonistic** – the action of one opposes the action of the other – in that one muscle must relax as the other contracts.

Muscle contraction

Muscle contractions are stimulated by nerve impulses. Each muscle consists of many fibres (see page 22). The fibres work independently of each other and are in contact with a nerve ending. A fibre cannot partly contract, it must be either relaxed or completely contracted. The muscle as a whole, however, may be composed of both relaxed and contracted fibres, the proportion of each varying according to the action that is to be produced. The contraction of muscle requires energy. This is derived from respiration of glucose in the blood and that stored in muscles as glycogen. The oxygen required for this metabolic process is provided by rich blood supplies in the muscle. If muscles are exerted for too long or too quickly, the amount of oxygen reaching them may become insufficient. In this event the muscles can continue to function in the absence of oxygen but only for a short while (see page 30). If this overexertion persists, an **oxygen debt** is created and lactic acid accumulates in the muscles causing **fatigue.** A period of rest is then required in order to re-establish the oxygen supply – first of all the lactic acid is oxidized, then more glycogen is converted to glucose and this metabolized to once again release energy by aerobic respiration.

The muscles are normally in a state of slight tension, which is called **'muscle tone'.** It keeps the structures to which the muscles are attached in their correct positions. Muscle tone does not produce fatigue because groups of muscle fibres alternate in their contraction and relaxation. Good **posture** is maintained if no group of muscles is unnecessarily tense or contracted in order to maintain balance or comfort. Bad posture results in fatigue through muscle strain. Standing and sitting should be as erect as possible for this permits the body weight to be distributed evenly upon the feet or buttocks. A curved back will put strain on the low back muscles and will also restrict adequate breathing movements. If the spine and shoulders curve forwards the abdominal contents are compressed leading to digestive problems.

When lifting a heavy weight, you should stand with both your feet apart to provide a stable base and bend your knees to avoid excessive force on one or two back muscles. The weightlifter obeys these rules. He uses the muscles of his arms and legs to raise the bar, rather than straining his back (see Figure 58).

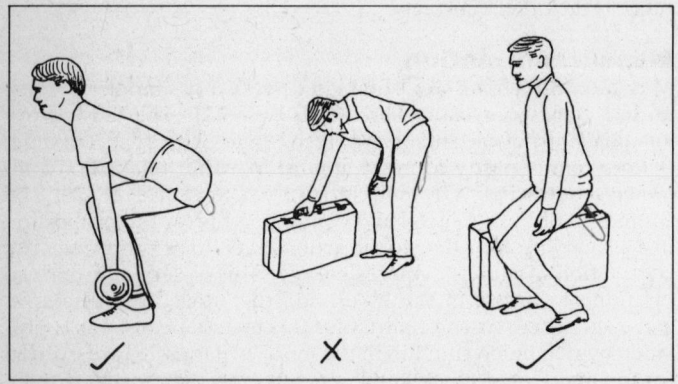

Figure 58. A suitcase should be lifted the same way as a weightlifter lifts his bar

Structure of bone

Bone is a living tissue. It consists of specialized cells called **osteoblasts** which are surrounded by an organic matrix. See Figure 59. The matrix is secreted by the osteoblasts and contains calcium salts, mainly **calcium phosphate.** These salt deposits give bone its hardness and strength. Osteoblasts are arranged in concentric rings around a nerve and blood vessel which run through a space in the bone known as an **Haversian canal.** Tiny channels called **canaliculi** run from the Haversian canal to the osteoblasts and contain capillaries. These bring tissue fluid to the living cells of the bone. The layers of bone laid down by the osteoblasts are known as **lamellae.**

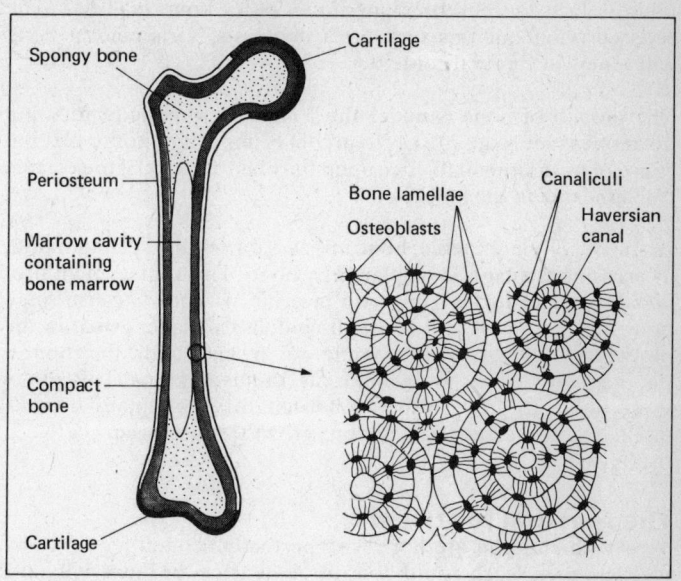

Figure 59. Structure of a long bone with a section through its compact bone region

Compact, or Haversian, bone tissue is found in the shafts of long bones, such as the femur and humerus. At the ends of these bones is another type of tissue, spongy bone, which is less hard and more stress-resistant. The outer layer of a long bone comprises a tough fibrous sheath called the **periosteum.** In the centre of a long bone is a cavity filled with a fluid called **bone marrow.** This is the site of production of red blood cells and granulocyte white cells (see page 74). It is now possible to transplant bone marrow from one person to another as a treatment for diseases such as **leukaemia,** although the recipient's immune system has to be suppressed by drugs so that the transplanted bone marrow cells are not rejected (see page 182). Leukaemia is a type of cancer which affects the white blood cells. It is usually fatal.

The growth of bone. Bones grow by deposition of insoluble calcium phosphate by the bone cells, the osteoblasts, a process called **ossification.** The osteoblasts produce an enzyme which converts blood-carried soluble phosphates into insoluble calcium phosphate. Within the cavity of a bone the reverse process is taking place thereby breaking down older bone material. The cells carrying out this task are osteoclasts. Gradually a bone will renew its mineral content.

The growth of bone is under the control of genetic factors and hormones (see page 118). Vitamin D is important in the diet because of its indirect influence upon the maintenance of the correct calcium level in the blood.

In the developing foetus, bone appears during the second month as areas of cartilage are replaced by bone. This replacement process is not finished by the time the child is born because small bones in the hands and feet still contain cartilage. Also, in the newborn ossification is not complete at the ends of the long bones, i.e. the femur, tibia, fibula, humerus, radius and ulna. This allows these bones to grow to their adult length. They finally become ossified at the cessation of all bone growth, which occurs between the ages of 18 and 25.

The feet and footwear

Nearly all children are born with perfectly formed feet. Yet by the time they reach adulthood, three-quarters of them will have developed deformities of the feet or poor stature caused by wearing ill-fitting shoes and socks. Not enough people realize that correctly fitting shoes in childhood is crucial. As explained above, the foot bones grow considerably in the first few years of life, so parents must regularly change their children's shoes. Socks and stretch-tights can similarly restrict the growing foot. Shoes should protect and provide support for the feet and keep them warm and dry. In wet weather, damp shoes (and other damp clothing) will lower the body's temperature because heat will be lost from the feet in drying out the shoes. Low body temperature reduces the child's resistance to infection, and coughs and colds result (see page 181).

High-heeled shoes throw too much of the body's weight forward on to the toes, upsetting the body's balance and posture. To compensate, the lumbar region of the child's back becomes more curved, causing back pain in later life and an inelegant gait.

Exercise, fatigue and rest
Physical exercise is important because it increases the capabilities of the body's vital tissues and organs, especially the heart, lungs, blood vessels and muscles. This helps to prevent the development of serious illnesses. Following sustained periods of exercise, muscles are able to function at reduced rates of energy expenditure. Exercise can also play a minor role in preventing obesity, although diet is a more important factor (see page 47).

Fatigue is a general weariness of mind or body associated with a decreased capacity to work. It is generally caused by metabolic waste products entering the bloodstream faster than they can be eliminated. **Yawning** is caused by acid waste products decreasing the pH of the blood. It mimics the effect of a high blood CO_2 level, thus stimulating the brain's respiratory centre to stimulate slow, deep inhalations and exhalations (i.e. yawns). If fatigue is to be avoided, periods of work should be followed by periods of rest. Monotonous tasks should also be avoided.

Fractures, dislocations, strains and sprains
As a result of a fall, blow or similar injury, musculo-skeletal damage may occur. If the patient complains of pain, inability to move a limb, or difficulty in breathing, a fracture or dislocation should be suspected. Fractured bone will produce pain near the bone and swelling and tenderness in the surrounding tissue. A simple fracture is any break which does not damage the tissue around it. A compound fracture is any break which extends to the skin. Because of this, bacteria can invade the body and inflame local tissue; much blood is usually lost. A greenstick fracture – an incomplete break of a bone, usually of the arm or leg – is most common in children, whose bones still have a considerable amount of cartilage. The usual injuries of a minor nature are **strains** (pulled muscles) or **sprains** (damaged ligaments).

Diseases of joints and muscles
Gout is a painful acute disorder of a joint brought about by the precipitation of uric acid crystals in the synovial capsule (see page 139). **Muscular dystrophy** is a defect of muscle fibres resulting in muscular wasting as though the nerve supply has been lost. **Arthritis** is a disorder of the joints where the cartilage linings at the ends of the bones degenerate; movement of these joints is particularly uncomfortable. It seems to be a result of 'wear and tear' and is thus more common in elderly people. **Rheumatoid arthritis** is a disease of the connective tissue in joints resulting

in inflammation of the fibrous capsule. The knuckle and wrist joints are the most commonly affected.

Patterns of growth

Physical growth of the body is usually measured by monitoring height and weight. The growth pattern generally follows a sigmoid curve (S-curve), i.e. one that starts slowly then increases to a maximum rate and reaches a peak after which no further changes occur (see page 158).

The first phase in the growth of an individual is marked by an increase in cell number, when hundreds of cells arise from a single zygote. The phase that follows is known as the 'grand period of growth', where rapid increase in the weight of the whole organism occurs. Growth is at a maximum in the foetus but continues at a high level during the first few weeks after birth and throughout the first year. From the age of two or three, the rate of growth slows down, becoming slowest of all just before the onset of puberty. **Puberty** (which occurs earlier in girls than boys) brings with it an **adolescent growth spurt** which continues until adult size is reached. At **maturity** the growth rate tails off and an equilibrium is reached between new cell growth and cell death. The final stage of life, **senescence,** is marked by decrease in weight as breakdown of tissues exceeds growth.

Development may proceed at different speeds, for instance the brain grows more quickly than the rest of the body in the first ten years, then slows down. The sexual organs hardly change at all during early childhood but rapidly increase in size at puberty.

Development of teeth

Teeth are part of the skeleton, being mostly made up of a substance similar to bone. There are two sets of teeth that develop in the human skull:
a. Temporary (milk) teeth – 20 teeth, in childhood.
b. Permanent teeth – 32 teeth, in adulthood.

The teeth are grouped according to shape, position and function. The **incisors** are chisel-shaped cutting teeth situated at the front of the upper and lower jaws. The **canines** are pointed grasping teeth situated next to the incisors. The third group, the **premolars,** are grinding teeth, as are the teeth at the back of the mouth, the **molars.**

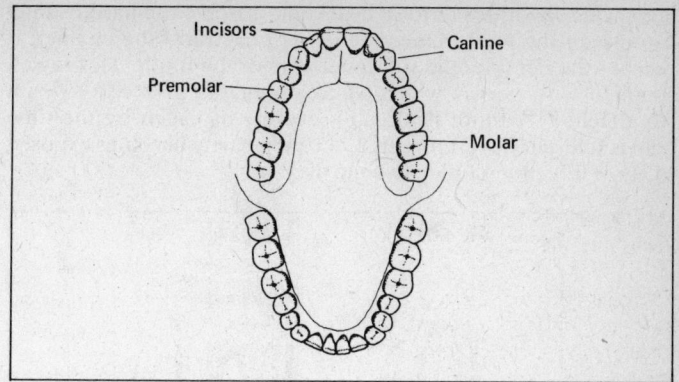

Figure 60. Dental arcade

The first teeth to appear in the child's mouth are usually the lower incisors, at the age of about six or seven months. The rest of the milk teeth emerge during the course of the following eighteen months or so. The permanent dentition begins at about six years when the first molars appear. The last **milk teeth,** the canines, are finally replaced by permanent canines at about twelve years. The third molars, the wisdom teeth, appear many years later, if at all.

The tooth consists of the crown, which is the portion above the gum. The neck is the region where the crown meets the root. The root is embedded in the jaw-bone. In cross-section, a tooth consists of three layers, the outer **enamel,** the **dentine,** and the inner **pulp,** which contains nerves and blood capillaries. In the root, instead of the enamel there is a **cement** layer (see Figure 61).

Tooth decay. Dental caries (dental decay) is predominantly a disease of Western civilizations. One major cause of the disease is eating refined carbohydrate foods (sugar, white flour and all foods containing these ingredients).

A thin layer of protein material derived from saliva is deposited normally on the teeth. If the teeth are not brushed regularly, this layer becomes impregnated with carbohydrates from the diet and micro-organisms present in the mouth to form a thick, sticky slime called **plaque.** In plaque, bacteria such as streptococci break down carbohydrates to form acids that dissolve the hard enamel coating on the teeth. Once the decay penetrates the enamel, it reaches the dentine, the soft inner core of the tooth. This is well supplied with nerves which when stimulated give rise to pain (toothache). A tooth is already severely damaged by the time pain is felt, but if left untreated, the pulp cavity becomes exposed to allow infection of the jawbone itself.

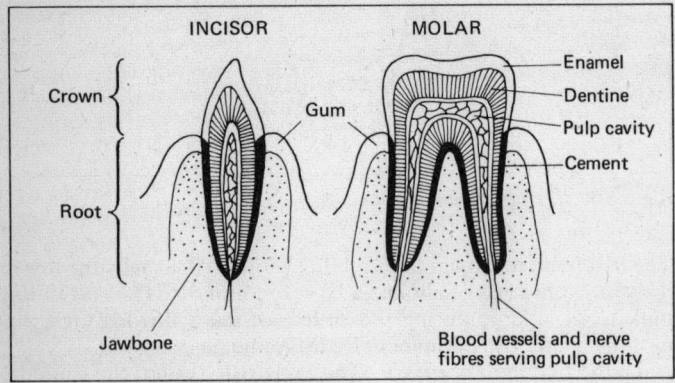

Figure 61. Structure of human teeth

Repair of a tooth damaged by caries consists of removing the decayed area using a drill and replacing it with a filling usually of silver and tin mixed with copper and zinc.

In England and Wales about 30 million fillings in decayed permanent teeth are carried out each year. A high proportion of these are carried out on children. Adding fluoride to drinking water has been shown to reduce tooth decay and hence the number of fillings required. But there is resistance to the mass **fluoridation** of water supplies. The safety and ethics of such mass medication are frequently heard arguments against this preventive dental health measure.

Key terms
Antagonistic muscles Two muscles situated either side of a joint, which oppose each other in their actions: on contraction, one muscle bends the joint, the other straightens it.
Appendicular skeleton The pectoral and pelvic girdles and the bones of the arms and legs.

Axial skeleton The skull, vertebral column, ribs and sternum.
Ball and socket joint Allows up to 360° of movement usually in several planes; the hip and shoulder joints.
Development Succession of stages in the life of an organism.
Differentiation Change in structure or function of the non-specialized type cell during development.
Endoskeleton Supporting material on the inside of the body; the skeleton of man.
Exoskeleton Skeleton on the outside of the body; e.g. the cuticle of insects.
Growth A permanent increase in size.
Hinge joint Allows up to 180° of movement in one plane only; the elbow and knee joints.
Ligament Tough, fibrous tissue linking one bone to another at a joint.
Joint Where two bones meet.
Synovial joint Freely movable, enclosed in a capsular ligament; cavity contains synovial fluid which reduces wear.
Tendon Tough, whitish, cord-like tissue linking a muscle to a bone.

Chapter 8

Reproduction and Inheritance

All living organisms have a finite lifespan, in other words they will eventually die. In fact death is the only certain event in a person's life. But although individual organisms always die off, species continue to survive because of the process of reproduction. Reproduction is different from growth. Growth is a process by which an individual increases in size, whereas reproduction is concerned with the creation of totally new, separate individuals. However, both processes involve the division of cells.

In Chapter 1 (page 31), we saw how human cells divide by a process called mitosis; one cell becoming two, which in turn become four cells and so on. This is essentially how the human body increases in size (grows) and replaces damaged or worn out cells. Even when you stop growing in height, mitosis is still actively going on in different parts of your body. Think of how damaged cells around a wound are soon replaced by new ones and how, for instance, in Britain a suntan quickly fades because tanned skin cells are continuously replaced by new white ones growing from beneath.

Many simple organisms such as unicells reproduce by mitosis. Each daughter cell resulting from this division becomes a new organism. Reproduction by mitosis is known as **asexual reproduction.**

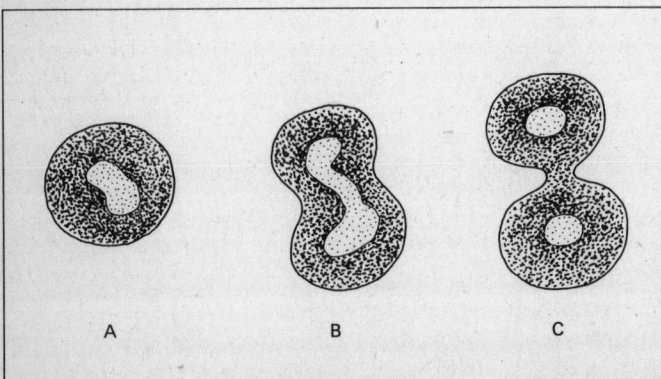

Figure 62. Asexual reproduction of a bacterium

Asexual reproduction is literally reproduction without sex. One organism simply multiplies to form two separate but identical organisms. By itself, an organism cannot produce two **different** offspring because only one set of genetic material (the chromosomes) is involved. This set of genetic material remains the same from generation unless affected by **mutation** (see page 176).

Man cannot reproduce by asexual reproduction, despite what you might read in a science fiction story. Man reproduces sexually, by having **sex.** Sex involves the joining together of a cell from one individual, the male, with a cell from another individual, the female. The resulting cell then grows into a new individual, quite separate and different from its parents. It is different because half its genetic material comes from one parent, and half from another. **Sexual reproduction** therefore allows organisms to vary from generation to generation (see page 167).

But human reproduction is more complicated than it sounds. First of all, the cells involved are not just ordinary body cells. They have to be special in one important respect: the number of chromosomes present in the nucleus. Man has 46 chromosomes in every cell. So if two of these cells joined together, the resulting cell would have 92 chromosomes. If that cell then grew into a new person, each of its cells would also have 92 chromosomes. If that person then reproduced with another person, the number of chromosomes in their offspring would be even greater (184). And so on.

So before two human sex cells can fuse to give rise to a new person, a reduction of their chromosome number must first take place. The process by which these special 'half-the-chromosome-number', or **haploid,** cells are produced is called **meiosis.** Cells with the normal number of chromosomes in their nucleus are called **diploid** cells. When two haploid cells fuse a diploid cell is produced.

There are two kinds of haploid cells. One, the **ovum** (plural ova), is found only in the female. The other, the **sperm** (plural sperm), is produced only by the male. Ova and sperm are sometimes known as **gametes** or germ cells. They are the only cells in the human body with 23 instead of 46 chromosomes in their nuclei.

Human reproduction involves the fusion of one sperm with one ovum to form a single cell, the **zygote.** The zygote is diploid and by mitosis divides to produce the new individual.

(a) FIRST MEIOTIC DIVISION

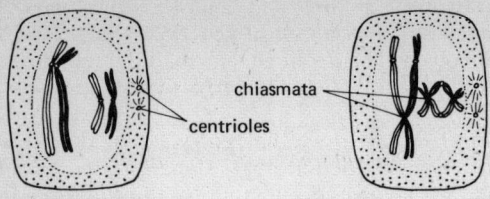

chiasmata

centrioles

(i) Pairs of homologous chromosomes lie close together. Each chromosome is already composed of a pair of chromatids

(ii) Homologous chromosomes separate except at the chiasmata. Chromatids exchange material by breaking and reforming at the chiasmata (crossing over)

 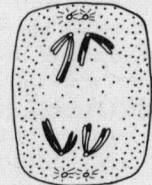

(iii) Nuclear membrane disappears. Spindle forms

(iv) Chromosomes migrate to ends of spindle, taking exchanged material with them

(b) SECOND MEIOTIC DIVISION

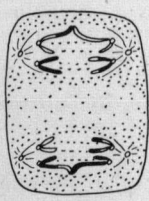 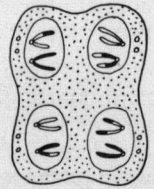

(i) The chromosomes divide again at right angles to the first division. This causes the chromatids to separate

(ii) Four new cells are formed, each containing the haploid number of chromosomes

Figure 63. Meiosis

The male reproductive organs and sperm production

Sperm are produced by the **testis,** of which each male has two. They lie in a protective extension of the abdominal cavity called the **scrotum** (or the scrotal sacs). This provides an environment that is 2°C cooler than the rest of the body. This lower temperature is essential for the development of the sperm. The testes also secrete the male sex hormone **testosterone** (see page 158).

The **epididymis** (plural epididymides) is a long coiled tube in which the sperm collect. A muscular sperm duct, or **vas deferens,** leads from each epididymis to a point just beneath the bladder, where it joins the **urethra.** This is a common passage for both sperm and urine from the bladder to the tip of the **penis.**

The penis transmits sperm from the male to the female. It consists of spongy tissue richly supplied with blood vessels. During **sexual intercourse,** the arteries in the penis dilate and blood fills the spongy tissue causing the penis to become stiff and erect. It can then be inserted into the vagina of the female.

The female reproductive organs

The ova are formed in the **ovaries,** of which there are also two. Very close to each ovary is an opening to a tube called the **oviduct** (or Fallopian tube). Ova pass down the oviducts into the **uterus** or womb. This is a thick-walled, muscular, hollow organ that is capable of considerable expansion. It is narrow at the lower end, forming a ring of muscle, the **cervix,** which separates the uterus from the **vagina.** The vagina is a muscular tube open to the outside at the **vulva,** which is enclosed by fleshy lips, or labia.

The urethra carries only urine from the bladder to the vulva.

Release of ova (ovulation). At puberty the ovaries begin to produce ova. One is produced every 28 days, the ovaries usually taking it in turn. Each ovum develops in a fluid-filled cellular structure called a **Graafian follicle.** Once fully formed in the nourishing medium of this follicle, the ovum bursts from the ovary into the body cavity. This process is called **ovulation.** The ovum is drawn into the funnel of the oviduct by the action of ciliated cells lining its walls, and continues along to the uterus. Meanwhile, the follicle cells in the ovary continue growing to produce a dark, solid mass of cells, the **corpus luteum.** This secretes the hormone **progesterone** (see later).

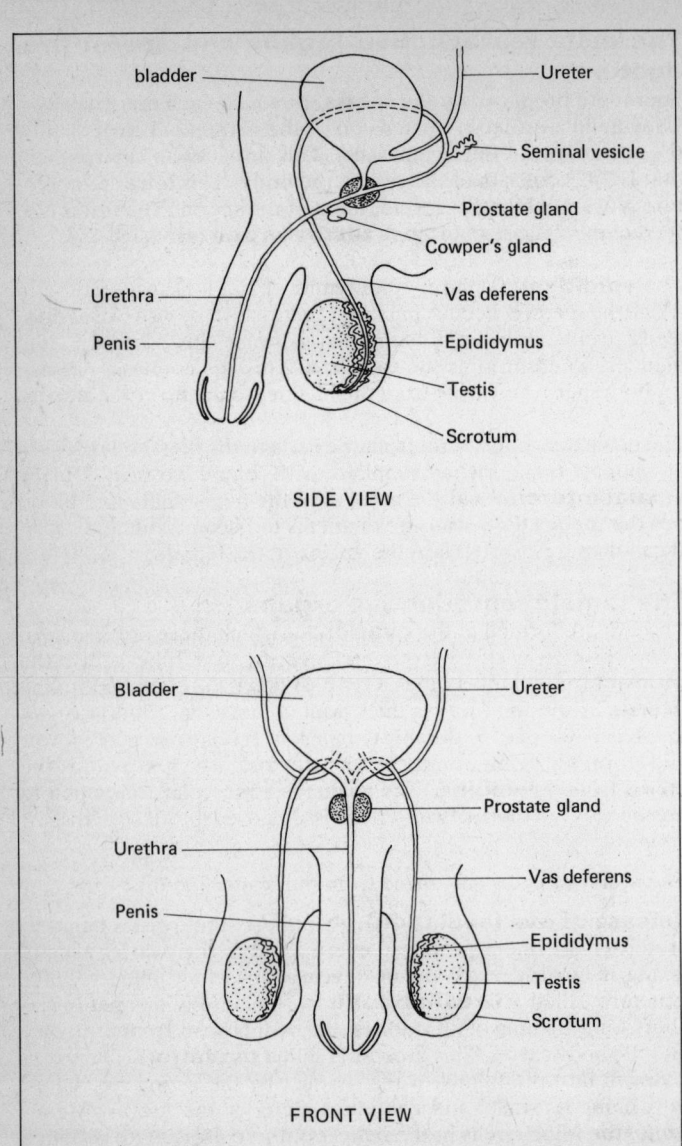

Figure 64. Male reproductive organs

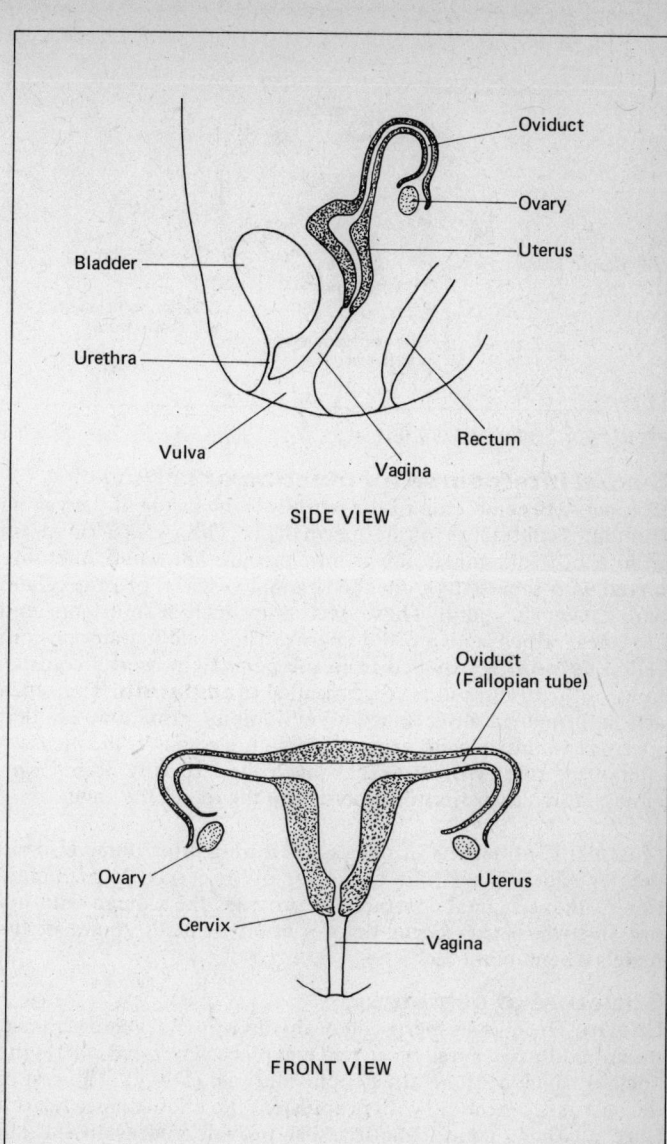

Figure 65. Female reproductive organs

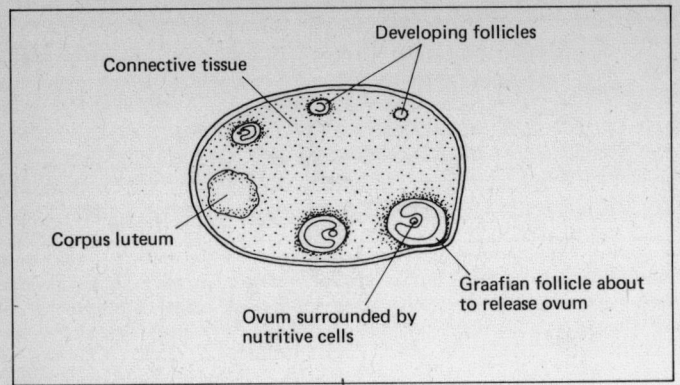

Figure 66. Section through ovary

Sexual intercourse (copulation or coitus)

Contact between the tip of the penis and the inside of the vagina stimulates contraction of the sperm ducts. This sweeps the sperm from the epididymides down into the urethra where they are mixed with secretions from the seminal vesicles, prostate gland and Cowper's gland. These secretions include nutrients and enzymes, which activate the sperm. The resulting suspension, called **semen,** is expelled from the penis by powerful contractions of the urethra in a process called **ejaculation.** The reflex action bringing about ejaculation inhibits urination, so that sperm is not mixed with urine. Ejaculation comes as the pleasant climax, or **orgasm,** of sexual intercourse, and it occurs with enough force to project the sperm into the top of the vagina.

Muscular contractions of the vagina during intercourse also aid the deposition of sperm in the region of the cervix. The culmination of these vaginal contractions provides the woman with her orgasm, which may occur before, after, or at the point of the male's ejaculation.

Structure of gametes

Sperm. The human sperm can be divided into five regions: head, neck, middle piece, tail piece and end piece. The head carries the nucleus which contains the genetic material (DNA). The rest of the sperm is concerned with propulsion: the middle piece bears a large number of mitochondria that provide energy to the tail piece, which is able to beat from side to side to enable the sperm to swim towards the ovum.

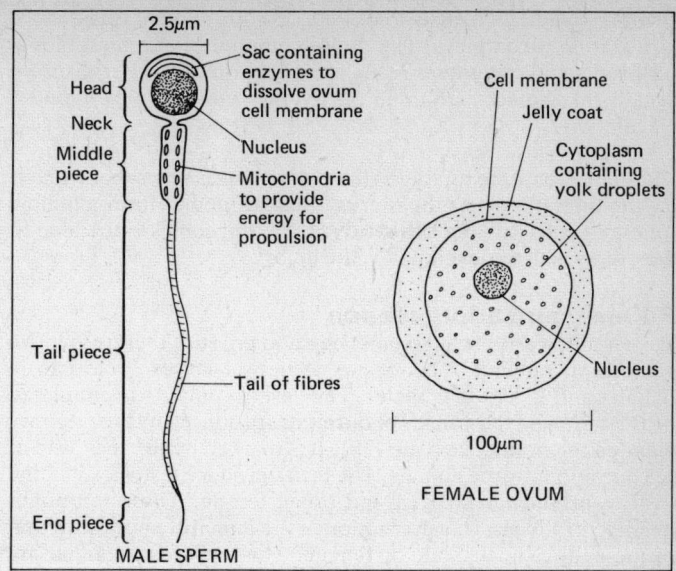

Figure 67. Human sperm and ovum

Ovum. The ovum is forty times larger than a sperm (diameter 100μm and $2.5\,\mu$m respectively) but it is much simpler in structure. The large nucleus, containing the genetic material, is surrounded by dense cytoplasm that in addition to the usual cell organelles also bears yolk droplets. These will provide a source of nourishment for the developing embryo. Around the outside of the ovum is a protective jelly coat consisting of glycoprotein.

Fertilization (conception). By undulating their tails, sperm can swim the length of the female genital tract. It is possible that they are aided on their journey by muscular contractions of the uterus but this effect is minimal. Only a small proportion of the sperm reach the ovum. There may be as many as 300 million sperm in a single ejaculation, whereas the female normally only sheds one ovum every 28 days.

Once a sperm reaches the ovum it releases a substance to assist its penetration of the ovum's jelly coat. Following this, further sperm are prevented from doing the same as a result of a chemical change at the surface of the ovum. The nuclei of the single sperm and the ovum then fuse to form a zygote and the process of fertilization is complete. The diploid number of chromosomes is restored and the zygote is ready for mitotic division. Fertilization usually takes place high up in the oviduct and within 24 hours of ejaculation.

Division takes place rapidly as the zygote moves down the oviduct. By the time it reaches the uterus it has divided to form a hollow sphere of cells called a **blastocyst.** This becomes embedded in the uterus wall to develop into the embryo.

Puberty and adolescence
Testes and ovaries (the gonads) begin to produce their respective gametes at puberty. This occurs at between ages 10 and 16 in females and 12 and 18 in males. The onset of puberty is stimulated by the action on the gonads of hormones produced by the pituitary gland in the brain (see page 118). It is marked by the secretion of sex hormones by the gonads. The testes produce testosterone, the ovaries produce oestrogen and progesterone. These hormones begin to produce a transformation of the immature body into that of the mature adult male or female. The body features that are altered by this process are known as the **secondary sexual characteristics.**

In females these are:
1. Growth of breasts and the female reproductive organs.
2. Growth of hair in the pubic area and under the armpits.
3. An interest in members of the opposite sex.
4. A general filling out of the body, with deposition of fat around the hips and an enlargement of the pelvis.
5. Onset of ovulation and menstruation (see below).

In males these are:
1. Growth of penis and body muscles.
2. Deepening of the voice.
3. Growth of pubic hair, hair in the armpits and on the face and chest.
4. An interest in members of the opposite sex.
5. Onset of sperm production in the testes.

Menstruation
From puberty until the onset of menopause (usually 45 to 55 years of age), a woman produces one (sometimes two) ovum every 28 days. Following ovulation, the uterus lining thickens and develops

an abundant blood supply so that should the ovum be fertilized it can readily **implant** in the uterus wall.

If implantation does not occur, the lining of the uterus is shed along with some blood. This cycle of ovum formation, ovulation, and thickening then degeneration of the uterus wall is known as the **menstrual,** or **oestrous, cycle** (see Figure 68).

For convenience, the cycle is considered to begin when blood is discharged from the uterus. This loss of blood and tissue fragments (days 1-4) represents the degeneration of the uterus lining and the unfertilized ovum. It is known as **menstruation** (or menstrual flow).

Once menstruation has ceased, the cycle starts again and a new ovum begins development (days 6-14) within a Graafian follicle. After ovulation (day 14), the empty follicle continues to grow to form the corpus luteum. In addition, the uterus lining becomes thickened and invaded with numerous blood vessels in preparation for implantation (days 14-24).

If implantation occurs, pregnancy follows (see page 162), but if it does not, the corpus luteum and uterus lining degenerate (days 24-28) and the menstrual period follows once again.

Hormonal control of the cycle. After menstruation, the pituitary gland in the brain secretes **follicle-stimulating hormone** (FSH). This stimulates both growth of the Graafian follicle and production by the ovary itself of the hormone oestrogen which in turn plays a part in the repair of the uterus lining following menstruation (see Figure 68).

Once a high level of oestrogen in the blood is reached, the pituitary gland is stimulated to produce **luteinizing hormone** (LH). This stimulates ovulation and the change of the Graafian follicle into the corpus luteum. The corpus luteum produces the hormone **progesterone** the principal function of which is to prepare the uterus lining for implantation. In addition, progesterone inhibits further production of LH. This has a feedback effect (see page 160) such that the corpus luteum stops producing progesterone. At this point menstruation takes place. The drop in progesterone level stimulates the production of FSH by the pituitary gland. The cycle is then repeated.

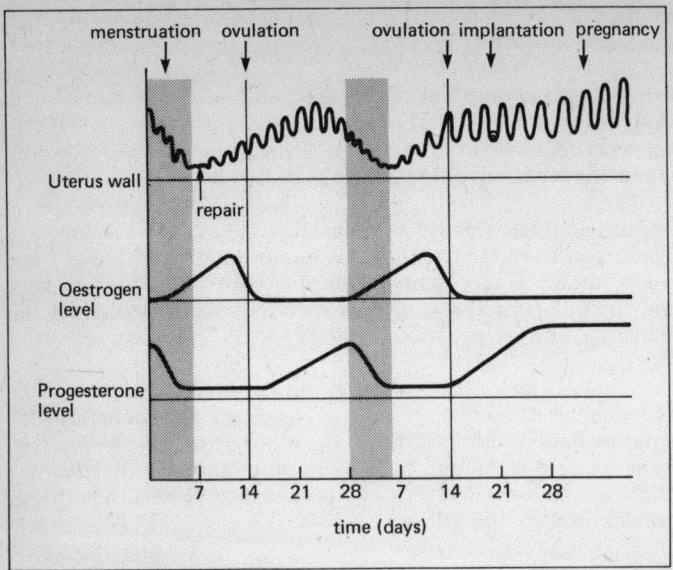

Figure 68. The oestrous cycle

Contraception: the prevention of pregnancy

Pregnancy may be prevented by either physical or chemical means or a combination of both. The main methods of contraception are as follows:

Coitus interruptus involves the male withdrawing his penis from the female before ejaculation. This is a very unreliable method of contraception and can lead to sexual disharmony and anxiety.

The rhythm method depends on accurate monitoring of the woman's oestrous cycle. Intercourse is restricted to the so-called 'safe period' either side of ovulation. The time of ovulation is often marked by a rise in body temperature so that by careful measurement of her temperature, a woman may be able to determine the day of ovulation and so avoid intercourse.

A condom, or **sheath,** fitted over the erect penis just prior to intercourse prevents the sperm from entering the female. Other such physical barriers consist of a rubber **diaphragm,** or **cap,**

fitted over the cervix of the female before intercourse. The effectiveness of such methods of contraception is increased by the additional use of chemicals in the form of a spermicidal foam or jelly that kills or immobilizes the sperm. Condoms are now available that are lubricated with spermicidal chemicals.

The **intra-uterine device (IUD),** another physical method of contraception, acts by preventing implantation of the embryo in the uterus. It is inserted into the uterus by a doctor and remains in place for up to three years. There are several different types.

The pill is a more reliable method of contraception. It contains synthetic oestrogen and progesterone, or progesterone only, which inhibit ovulation. As no ovum is normally produced the risk of pregnancy is minimal.

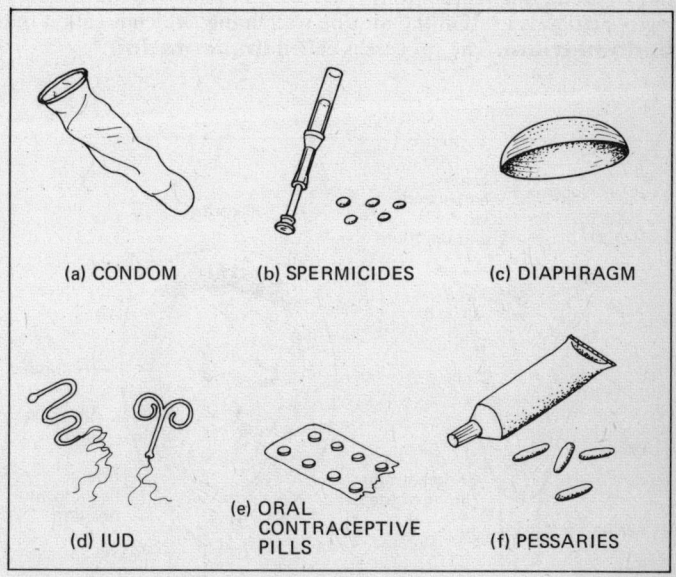

Figure 69. A range of contraceptives

All methods of contraception provide a temporary and reversible means of birth control. However, for men or women who no longer want children, sterilization is becoming a popular option.

In women this involves the cutting and tying of the oviducts, and in men the cutting of the sperm ducts. These methods are extremely effective, but cannot be easily reversed.

A new and more imaginative method of contraception is becoming popular in the United States. Men are depositing semen in a sperm bank and then having a vasectomy. When later they want a baby, they return to the bank, withdraw their 'deposit' and use it to artificially inseminate their wives. Unfortunately the storage of sperm requires it to be kept at very low temperatures (about $-250°C$) using liquid nitrogen as a coolant. Recently one of these cooling plants broke down and irate depositors sued the sperm bank after their immortality literally melted away.

Pregnancy

Pregnancy commences when the zygote begins to divide and passes along the oviduct to the uterus. In the uterus the dividing zygote embeds itself in the uterine wall lining, which is called the **endometrium.** The process is called **implantation.**

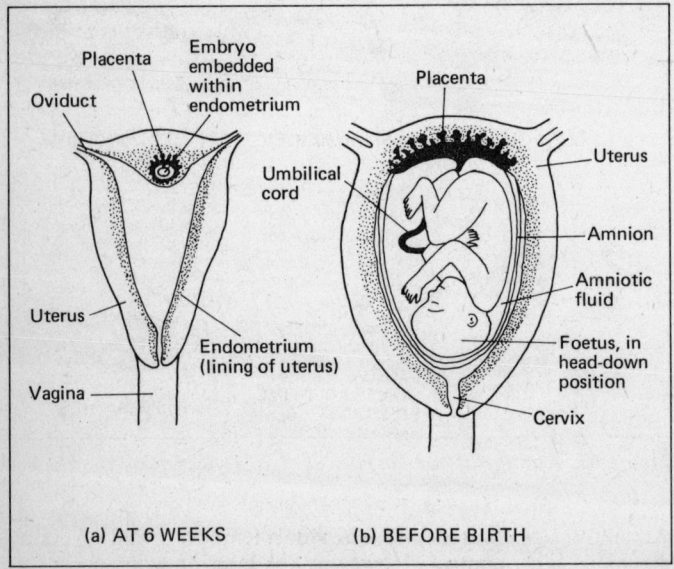

Figure 70. The foetus in the uterus

The rapidly dividing mass of cells is now called the embryo. As development proceeds, the **embryo** becomes surrounded by protective membranes. The **amnion** is the innermost membrane and encloses a fluid filled cavity in which the embryo is suspended. The fluid protects and supports the embryo while allowing it freedom of movement. In the region of the uterine wall, the two other membranes form the **placenta** and this connects with the abdominal region of the embryo via the **umbilical cord**. By about the eighth week of pregnancy the embryo is well developed and becomes known as a **foetus.**

The umbilical cord contains two arteries carrying deoxygenated blood from the foetus to the placenta and a vein that returns oxygenated blood to the foetus.

The placenta is an organ that plays a vital role throughout pregnancy. It secretes hormones which maintain pregnancy and conducts nutrients (food and oxygen) from the maternal blood vessels to the foetal blood vessels and foetal wastes (carbon dioxide and urea) in the opposite direction. It also acts as a protective barrier, preventing harmful and toxic substances from entering the foetal blood system. This is possible because the foetal and maternal blood circulations remain separate. However, many drugs, such as sleeping pills (e.g. thalidomide), are able to cross the placenta/embryo barrier and adversely affect the development of the child, as does the German measles (rubella) virus, which damages the foetus's nervous system. Smoking and alcohol consumption are also inadvisable during pregnancy as they can affect the growth of, or even damage, the foetus.

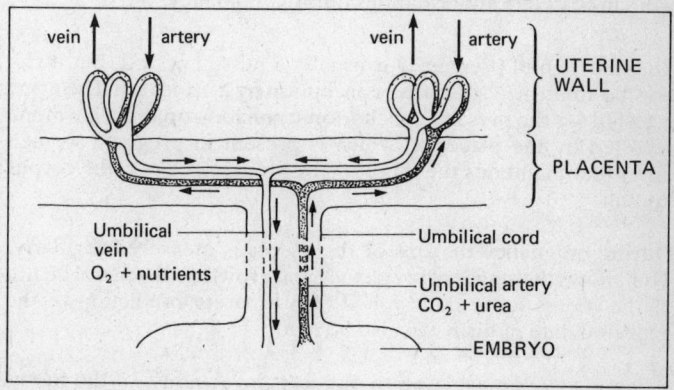

Figure 71. The foetal circulation

The structure of the placenta is adapted to these functions as it has a large surface area that is richly supplied with blood. Also, the membranes separating the foetal capillaries from the maternal circulation are very thin, allowing maximum exchange of substances carried in the blood systems.

Hormonal control during pregnancy. If after ovulation fertilization occurs, the oestrus cycle is interrupted. The corpus luteum persists and continues to secrete progesterone and the ovary itself continues to secrete oestrogen. Both these hormones maintain the development of the endometrium and progesterone prevents further ovarian follicles from developing during pregnancy (see Figure 68).

After three or four months of pregnancy the placenta largely takes over the function of secreting oestrogen and progesterone. The hormones are then responsible for the enlargement of the milk-secreting mammary glands (breasts). After birth, milk production, or **lactation,** is promoted by the hormone **prolactin,** which is secreted from the pituitary gland. **Oxytocin,** an additional hormone secreted by the pituitary gland, causes uterine muscle to contract at the end of pregnancy; which is usually 38 weeks after conception. This hormone is sometimes given to the mother to induce birth when pregnancy has lasted too long or the foetus is believed to be getting too large for birth to take place smoothly.

Prenatal screening. This consists of regular monitoring of the woman to detect abnormalities during pregnancy.

The first sign of pregnancy is usually a missed period, but it can now be confirmed soon after conception by a urine test. The urine is tested for the presence of chorionic gonadotrophin, a hormone secreted by the placenta (which is present in pregnant women only) that maintains the secretion of progesterone by the corpus luteum.

During pregnancy the size of the uterus is measured regularly. This allows the stages of pregnancy and growth and general health of the foetus to be determined. An accurate prediction of the expected date of birth can also be made.

X-rays are no longer used to monitor the progress of the foetus on account of their possible harmful effects, and have been superceded by **ultrasound,** in which sound waves are used to scan the uterus. Further tests may be made by puncturing the amniotic membrane to obtain amniotic fluid and foetal cells, a technique known as **amniocentesis.** Examination of the cells will reveal the sex of the foetus and possibly any abnormalities in its chromosomes. Chromosome defects such as those found in Down's syndrome (mongolism) may be detected by this test. Spina bifida, which is a serious abnormality of the central nervous system, may also be detected by the method.

Birth (parturition). After approximately nine months from fertilization, the so-called gestation period draws to an end and the foetus is ready to be born.

The initial stage of birth involves movement of the foetus until its head lies above the cervix, followed by rhythmic contractions of the uterine wall. This marks the onset of **labour** during which the muscular contractions of the uterus become stronger and more frequent and the amnion breaks releasing the amniotic fluid ('breaking of the waters').

With powerful contractions of the involuntary muscles of the uterus combined with voluntary contractions of the abdomen, the foetus is forced out of the uterus through the expanded cervix and vagina. The bones in the foetal skull are able to slide over each other a little and so ease passage through the mother's pelvic girdle. In addition, in women the cartilage joint at the front of the pelvis is able to open slightly.

On contact with the outside world, the baby takes its first breath. At about this time the umbilical pulse ceases and the umbilical arteries constrict. The umbilical cord is tied off about 5 cm from the baby and again near to the mother. The cord is cut between these two ligatures. Once this separation from the mother has taken place, the baby begins to rely on its own blood circulation. Meanwhile, the placenta becomes detached from the uterus wall and is later expelled (as the afterbirth) following further contractions of the uterus.

A new life. While the baby was inside its mother, it was kept warm, fed and supplied with oxygen. All its waste products were removed by the mother's blood circulation. But as soon as it is born, it must begin to fend for itself, learning to eat and excreting its waste products as an independent being.

The newborn baby's first cries open up the millions of tiny airways inside its lungs. From now on it will breathe for itself. Food is obtained as milk from the mother's breasts. Human milk differs considerably from cow's milk and formula (artificial) milk in the proportions of its main ingredients. For instance, human milk contains twice as much sugar as cow's milk and half as much protein. However, for the first few days after birth the mother's breasts secrete a special kind of milk called **colostrum.** This contains four times more protein and less fat and sugar than ordinary mother's milk. The mother produces colostrum at her own body's expense, even if she is undernourished, as it is important the baby receives it. Later on, if the mother is underfed or in poor health, her milk secretions will dry up; the baby must then be given sterilized cow's milk or formula milk.

A baby should be fed on its mother's milk for as long as possible not only because it is a perfect food and feeding at the breast helps to form a bond between mother and child, but also because the milk contains antibodies that will protect the baby from infection. A baby's own defences against micro-organisms do not develop fully until the end of its first year (see Chapter 9).

As a child grows up it learns to become less and less dependent on its mother and father. But during its first years of life it needs a secure bond of attachment to both parents. By the age of four it has learnt all the rules of language from its parents, and with their help learnt to move around and explore its environment. At this time it is also important that it receives a balanced diet and one rich in protein so that its body can grow normally (see page 14).

Abortion. Almost 200,000 legal abortions (also called terminations of pregnancy) are performed in Britain each year. Often this is because the developing foetus has a defect of growth, or **congenital abnormality,** which has been detected during pregnancy. In other cases, abortions are carried out when the continuance of the pregnancy would seriously risk the life of the mother or injure her mentally or physically.

An early pregnancy is terminated by removing the embryo by means of suction using a flexible plastic tube which is passed into the uterus through the cervix. This method is very quick and can be performed using only local anaesthetic. Later pregnancies are aborted by artificially inducing contractions of the womb using a hormone, or alternatively the foetus can be removed surgically by **Caesarian section,** which involves cutting the mother's abdomen and uterus. Occasionally births are also carried out this way.

Heredity

Children resemble their parents but they are also different from them in many respects. The study of the similarities and differences between parents and their offspring is known as **heredity.** You should already realize that a child develops from a zygote which is formed from the union of two gametes: a sperm from the father and an ovum from the mother. All the information that is needed for the development of the child is contained in that single-celled zygote. It exists in the chromosomes, the strands of DNA that are found in the nucleus.

The two gametes had 23 chromosomes. The zygote, and every human body cell, has 46 chromosomes. So the zygote receives half its chromosomes from the mother and half from the father. Because it is the expression of the information within the chromosomes that determines the characteristics of the individual, it should not be surprising that the child, and later the adult, developing from the zygote resembles both mother and father.

But genetics is not as simple as this. Why should some of a child's features resemble those of his father and mother and others not? How is it that some children do not seem to resemble their parents at all? And of those children born as twins, why in some cases are the individuals identical to one another and in other cases quite unalike? The science of genetics can explain a lot of these phenomena and the starting place of the subject is the chromosome.

A chromosome is a long strand of **DNA** (deoxyribonucleic acid) (see page 15). Chromosomes can be divided up into regions called **genes,** each gene being responsible for the production of a particular protein and hence a particular characteristic.

How DNA works. Proteins are assembled on **ribosomes** in the cell cytoplasm. So for DNA to control the assembly of proteins it must migrate out of the nucleus to the ribosomes or exert its control through an intermediary substance. Because DNA is never detectable in the cytoplasm, an intermediary substance must be used. This substance is a type of RNA (ribonucleic acid) called **messenger RNA,** or m-RNA.

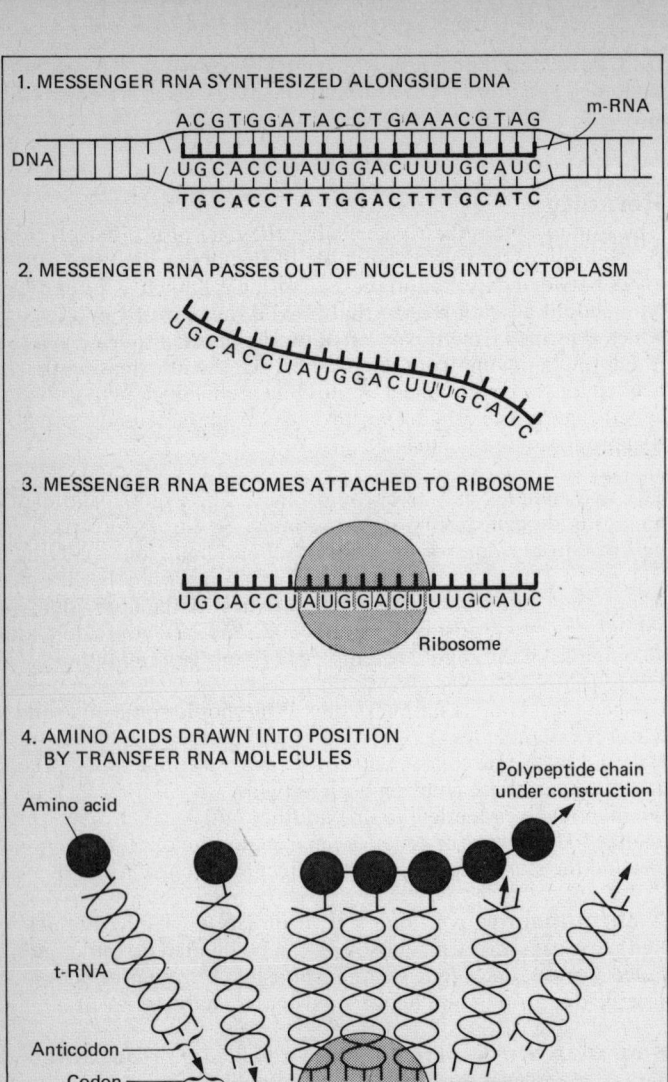

Figure 72. How proteins are assembled using the DNA code

Strands of m-RNA are synthesized alongside DNA strands in the nucleus (see Figure 72). Remember that RNA consists of one strand instead of two; contains the sugar **ribose** rather than **deoxyribose**; and contains the base **uracil** instead of **thymine.** The DNA serves as a **template** for the formation of the m-RNA. The DNA unzips in the appropriate place and free RNA nucleotides align themselves opposite one of its two strands. **Cytosine** always pairs with **guanine** and **adenine** with **uracil.** This ensures that the sequence of bases on the DNA molecule (the **genetic code**) is accurately transferred to the m-RNA molecule. Once assembled, the m-RNA peels off the DNA template and leaves the nucleus through the pores in the nuclear membrane.

In the cytoplasm, the m-RNA becomes attached to a ribosome. Here it causes the assembly of proteins through the use of yet another intermediate substance called **transfer RNA,** or t-RNA. There is one type of t-RNA for each type of amino acid. A t-RNA molecule is a short single-stranded piece of m-RNA but the strand double-backs on itself and is twisted into a helix. One part of the t-RNA molecule has three unpaired bases projecting from it. These three bases are called the **anticodon** and the constituent nucleotide bases vary according to which amino acid the t-RNA carries. Anticodons correspond to the series of three bases (called **codons**) on the m-RNA. Thus t-RNA molecules are drawn to the m-RNA according to the order of codons. Once they are aligned on the m-RNA, peptide bonds are formed between the amino acids and a protein is formed. This sequence of events is shown in Figure 72.

Genes. A good example of a human gene responsible for a single, clearly defined characteristic is that producing the shape of the earlobe. There is a gene producing lobes of normal shape. Everyone has this gene on one of their chromosomes but in a few people the gene is slightly different, resulting in the production of unlobed ears (see Figure 73).

In fact, the gene for earlobes exists on more than one chromosome. Remember that in the nucleus each chromosome has a partner of the same size. The gene for earlobes exists on this partner (or **homologous**) chromosome as well. The same is true for every other gene, with one or two exceptions that are explained later. But although each gene exists on two chromosomes, the form of the gene need not be the same on both. For example, the earlobe gene produces normal or unlobed ears. These different forms of the same gene are known as **alleles.** A person can have the allele for normal earlobes on one chromosome and the allele for no earlobes on the other or any combination of the two.

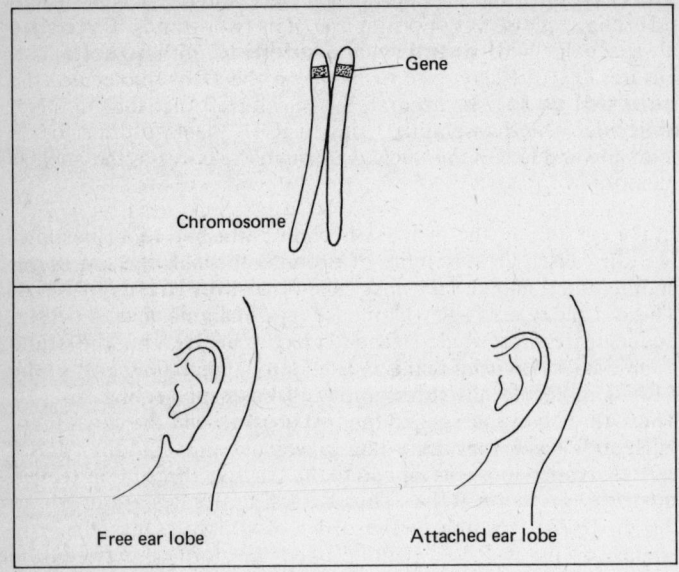

Figure 73. A gene and its effect

In genetics, alleles are represented by letters. So in this example let:

E = the allele for normal earlobes
e = the allele for no earlobes

Three combinations of these alleles are possible: EE, Ee and ee. An individual's specific allele combination for a characteristic is known as his **genotype.** Now there are only two possible physical expressions of these alleles, that is a person has normal earlobes or no earlobes. These two possibilities are known as **phenotypes.**

To recap, the genotype is what is present chemically (in the form of the DNA code) on the genes, the phenotype is the outward expression of the genotype.

The characteristic resulting from expression of the genotype depends upon the relationship between the two alleles. In most cases one allele is able to exert a dominant effect upon the other allele, and so its characteristic type forms the phenotype. This is the case with earlobes. E is dominant to e, so the genotype Ee will produce normal earlobes. E is known as the **dominant allele,** e as the **recessive allele.**

On the basis that the prefix *homo* means same, and the prefix *hetero* means different:

an individual who is EE is known as **homozygous dominant.**
an individual who is Ee is known as **heterozygous.**
an individual who is ee is known as **homozygous recessive.**

In the case of earlobes, only those who are homozygous recessive (ee) will have no earlobes.

Consider what happens if two homozygote parents (one EE, the other ee) produce a family.

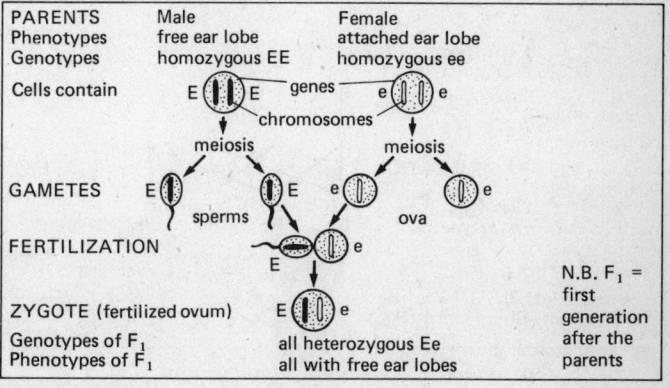

Figure 74. Genetic cross: Inheritance of earlobes

Each child has earlobes in that the dominant allele, E, is present in all zygotes.

The same dominance relationship exists between the alleles responsible for producing the Rhesus (Rh$^+$) factor protein in the blood, mentioned in Chapter 4. This protein, which occurs on the wall of erythrocytes, is present in some people but not others. The conditions are referred to as Rhesus positive (Rh^+) and Rhesus negative (Rh^-) respectively. The allele Rh^+ (appearance of protein) is dominant to the allele Rh^- (absence of protein).

If a homozygous Rh^+ woman marries a Rh^- man, their offspring will all be Rhesus positive but possess the recessive allele in their genotypes (see Figure 75).

If a heterozygous Rh^+ male and female marry, their offspring will be in the ratio of three Rh^+ to one Rh^- phenotype (see Figure 75).

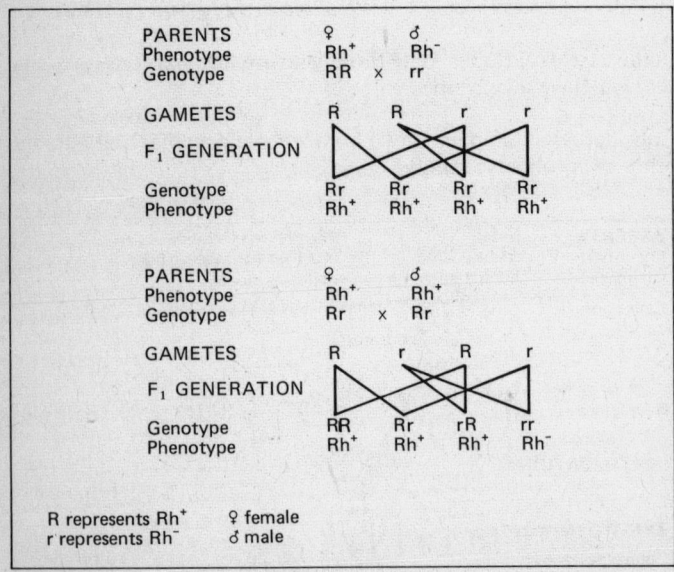

Figure 75. *Inheritance of the Rhesus blood group*

Incomplete dominance. Some alleles do not show the dominance effect. This occurs with the genes that determine the ABO blood group system. Everyone can be classified as having one of four major blood groups: group A, group B, group AB, or group O (see page 80). In other words, there are four phenotypes. The ABO gene exists as three alleles: A, B and O.

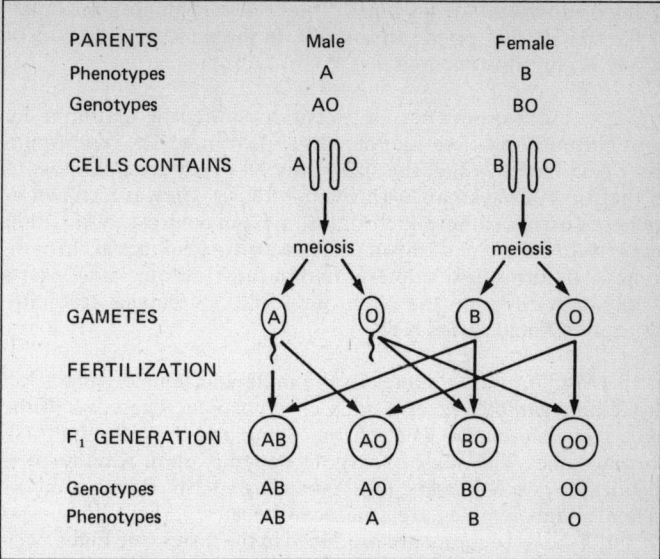

Figure 76. Inheritance of ABO blood groups

Allele A produces antigen A on the red blood cells and allele B produces antigen B. The O allele produces no antigens, and is recessive to A and B. However, the A and B alleles always express themselves completely so when together produce antigens A and B on the red cells. They are referred to as co-dominant alleles and the situation they give rise to as incomplete dominance..

Thus for the four possible phenotypes there are six genotypes:
Phenotype   O   A   B   AB
Genotypes   OO   AA AO   BB BO   AB

Remember both A and B are dominant to O.

We can examine what happens where a heterozygous blood group A person mates with a heterozygous blood group B person (see Figure 76).

Sex determination. In every human cell there are 23 pairs of chromosomes in the nucleus. Of these, 22 are homologous chromosome pairs and the remaining pair comprises the **sex**

chromosomes. In the female these are an homologous pair of the so-called **X chromosomes.** In the male the pair consist of one **X** chromosome and one **Y** chromosome.

The sex chromosomes are an exception to the rule that homologous chromosomes are identical in appearance: The Y chromosome is in fact only half the size of the X. (They are homologous in that they always pair with one another.) They are known as **heterosomes** (different chromosomes) in contrast to the other pairs of chromosomes known as **autosomes** (identical chromosomes). In terms of sex determination the Y chromosome exerts a dominant effect on the X chromosome, so females are genotypically XX and males XY.

In the formation of gametes in the male and female gonads, all the female gametes receive an X chromosome, whereas half the male gametes receive an X chromosome and the other half a Y chromosome. The sex of a zygote depends upon whether it is fertilized by an X-bearing or a Y-bearing sperm. Equal numbers of males and females are produced because equal numbers of X- and Y-bearing sperm are produced in the testes (see Figure 77).

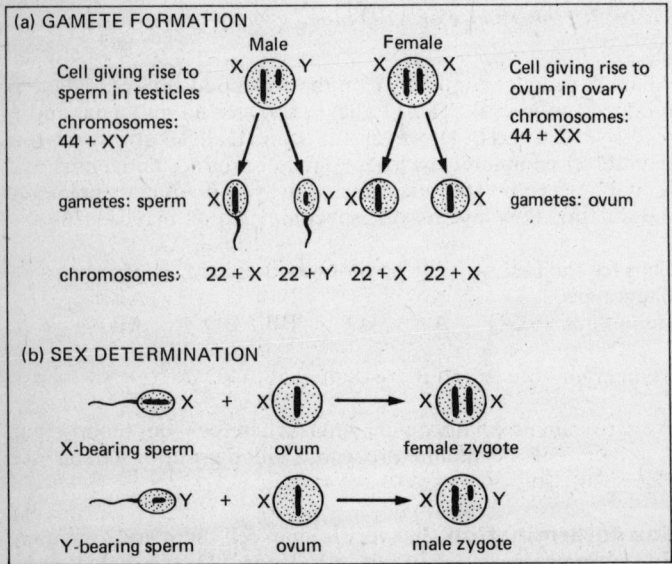

Figure 77. Sex determination

Sex-linked genes. Genes carried on the X and Y chromosomes must clearly be transmitted with those determining sex. They are said to be sex-linked.

In the male, the sex chromosomes are not of equal size and so any gene on the unpaired part of the X chromosome will not be masked by a corresponding allele. This is true whether the gene in question is recessive or dominant.

In the female, the same gene will be present on both the X chromosomes. A recessive gene will not now be likely to be expressed as it can be matched by a dominant gene on the corresponding homologous chromosome. The female in this case can carry the recessive gene but will not outwardly show the corresponding characteristic. She is referred to as a **carrier.**

An example of a sex-linked recessive gene (i.e. one that is expressed in the male and carried by the female) is that for red-green colour blindness. An affected individual cannot distinguish between reds and greens. A colour-blind man married to a normal sighted woman transmits his colour-blind gene to his daughters. Since they are heterozygous, they will not be colour-blind but will be carriers of the recessive gene (see Figure 78).

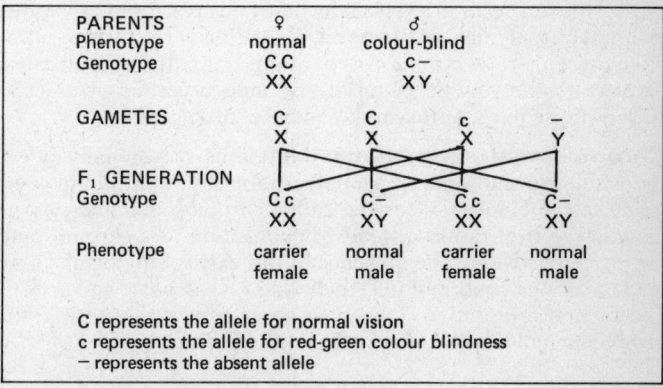

Figure 78. Inheritance of colour blindness

If one of the daughters marries a normal-sighted man and produces a son, there is a 50 per cent probability that he will be colour-blind. A daughter from the marriage will be phenotypically normal but there is a 50 per cent probability that she too will be a carrier (see Figure 78). Red-green colour blindness is more common in men than in women, who can only inherit the disorder if their father is affected and their mother is a carrier.

An additional example of sex-linked genes is provided by the gene for haemophilia. This is a condition in which blood takes an abnormally long time to clot, resulting in excessive bleeding from any wound. The gene responsible is usually passed on by female carriers and expressed phenotypically in her male offspring.

Mutations. A mutation is a spontaneous and permanent change in a gene usually due to an alteration of the nucleotide base sequence in the DNA which comprises the genetic code. This can result in an alteration in the phenotype unless, that is, the mutation is in an allele the effect of which is masked by a more dominant allele. Many human disorders are due to the effects of mutant genes. For example, **phenylketonuria,** a rare cause of mental deficiency, is due to a mutant gene that fails to code for the manufacture of an essential metabolic enzyme. The result is that the body cannot metabolize one amino acid, phenylalanine, the partial breakdown products of which collect in the urine. These abnormal partial breakdown products act directly on the brain producing mental retardation and disability.

Mutations can occur in a cell at any time and are caused by various external factors called **mutagens.** Radiation in the form of ultra-violet light, X-rays, cosmic rays or atomic radiation may all cause damage to the genetic material and indeed certain chemicals, such as those in cigarette smoke, may be 'mutagenic'.

Chromosomal aberrations. Mutations occasionally affect the whole chromosome. Such defects are often, but not always, lethal to the embryo. These so-called chromosomal aberrations include the triplication instead of duplication of a chromosome during cell division. The presence of this extra chromosome may lead to serious abnormalities, including a weak heart and growth and mental retardation. One such aberration is responsible for producing individuals suffering from mongolism.

Key terms
Abortion Termination of pregnancy.
Allele One of a number of forms of a single gene.
Asexual reproduction Reproduction without gametes, by simple cell division (mitosis).

Contraception Method of preventing pregnancy.
Copulation Sexual intercourse.
Corpus luteum Part of a Graafian follicle in the ovary that after ovulation produces hormones regulating pregnancy.
Diploid The normal number (full complement) of chromosomes in a cell, in man 46.
Dominant allele An allele which is always expressed in the phenotype of the organism.
Ejaculation Passing of sperm out of the penis.
Embryo Early stage in development of a baby in the uterus, produced by division of the zygote.
F_1 generation The first filial generation; the first generation of offspring of a genetic cross.
Fertilization Fusion of a male and a female gamete (sperm and egg).
Foetus Stage of development of an unborn baby at which most organs are well-formed; about 8-10 weeks after fertilization.
Follicle A fluid-filled space in the ovary containing an ovum.
Gametes Sex cells, i.e. sperm and ova.
Gene A portion of a chromosome which controls the synthesis of a protein.
Genotype The genetic make-up of an organism.
Genetics The scientific study of genes and the way they control characteristics.
Haploid Half the normal (diploid) number of chromosomes, in man 23.
Heredity The inheritance of characteristics from one's parents.
Hermaphrodite An organism which has both male and female sex organs.
Heterozygous An organism containing two different alleles of a particular gene.
Homozygous An organism containing two identical alleles of a particular gene; true-breeding.
Implantation The process whereby a zygote becomes embedded in the uterus wall.
Incomplete dominance Where two alleles of a gene are both expressed in the phenotype, i.e. co-dominance.
Intra-uterine device (IUD) A contraceptive device inserted into the uterus.
Lactation Production of milk by the mammary glands.
Meiosis Cell division to produce gametes; involves reduction by half of the diploid number of chromosomes.
Menopause Age at which women lose their reproductive ability.
Menstruation Sloughing off of the lining of the uterus, occurring in adult women every 28 days.

Mutation A spontaneous change in a chromosome.
Oestrogen Female sex hormone.
Ovum Female sex cell produced by the ovary.
Ovulation The release of an ovum from the ovary.
Parturition Birth.
Phenotype All visible (internal and external) features of an organism.
Placenta The organ through which a developing embryo obtains food and oxygen from its mother and eliminates waste products into its mother's blood circulation.
Pregnancy The period of 9 months during which a mother carries an unborn baby in her uterus; from fertilization of the ovum to birth.
Prenatal screening Monitoring a pregnant woman to detect abnormalities in the unborn baby.
Progesterone Hormone produced by the corpus luteum and by the placenta during pregnancy which primarily maintains the state of the uterine wall.
Puberty Sexual maturity.
Recessive allele An allele which is not usually expressed in the phenotype (due to the presence of its dominant partner).
Semen Nutrient fluid produced by the testes and associated glands which accompanies sperm.
Sex-linked gene A gene which is located on a sex chromosome (usually the X chromosome).
Sexual reproduction Reproduction involving the fusion of the nuclei of gametes; usually involves two individuals, the male and the female.
Umbilical cord A tube containing blood vessels that connect a mother and her developing embryo.
Uterus Pear-shaped, muscular organ which can massively enlarge; bears the developing foetus.
Zygote The diploid cell which results from the fusion of a sperm with an ovum.

Chapter 9
Health and Disease

A person's body works normally most of the time, with physiological factors remaining in a state of equilibrium. This is generally referred to as health. But occasionally the equilibrium is disturbed either temporarily or permanently, and that person becomes ill and/or disabled. A **disability** is a permanent abnormal functioning of the body which, like an illness, can be either **physical** or **mental.** Disabilities were in the past described as handicaps, but the name was changed because many people do not feel handicapped by their disability, This is particularly true of physically disabled people who, because of paralysis, live their lives in a wheelchair but are able to function normally in almost every respect.

A physical or mental illness can last for a variable period of time. An **acute** illness is usually short and sharp, and the patient either recovers quite quickly in a matter of days or weeks, or dies. A **chronic** illness is less life-threatening but can last for many years. Most acute illnesses, such as influenza or whooping cough, are infections caused by micro-organisms. Chronic illnesses, on the other hand, tend to be caused by a gradual degeneration, or deterioration, of a part of the body, an example being heart disease or bronchitis. A chronic illness (or disease) differs from a disability in that a patient can eventually be cured of or recover from it.

When people become ill they experience **symptoms,** which are the unpleasant sensations, such as pain or nausea (feeling sick), that prevent them from carrying out their normal daily activities. Some people have only mild symptoms and are able to lead their usual life; they are not usually described as being ill. Illness and injuries produce **signs** as well as symptoms. The signs of an illness are what a doctor looks for when doing a medical examination. A skin rash, a high temperature, loss of weight, diarrhoea, sugar in the urine and a furred tongue are all signs. Symptoms are the effects of an illness which are experienced by the individual and cannot usually be observed by a doctor. All diseases exhibit certain signs and symptoms which a doctor is trained to recognize.

Diseases can be either transmissible or non-transmissible. Transmissible, or infectious, diseases can be passed on from person to person. Non-transmissible diseases cannot.

Type of Disease	Examples
Transmissible	
1. Diseases caused by micro-organisms	influenza, cholera, whooping cough, measles, mumps
Non-transmissible	
2. Metabolic disorders	diabetes, phenylketonuria
3. Nutritional deficiency diseases	kwashiorkor, beri-beri, rickets, scurvy
4. Degenerative disease	arthritis, coronary heart disease
5. Cancer	breast cancer, leukaemia
6. Mental illness	depression, schizophrenia

Transmissible diseases are caused by **pathogenic micro-organisms,** which are able to invade and disrupt the body. There are four main types:

1. Bacteria. These are single-celled organisms which are to be found everywhere, from the sea bed to the upper atmosphere. Only a small proportion of them are pathogens (cause diseases); many perform useful functions. For example, some live in the human gut and help in digestion. Others are used by man in the

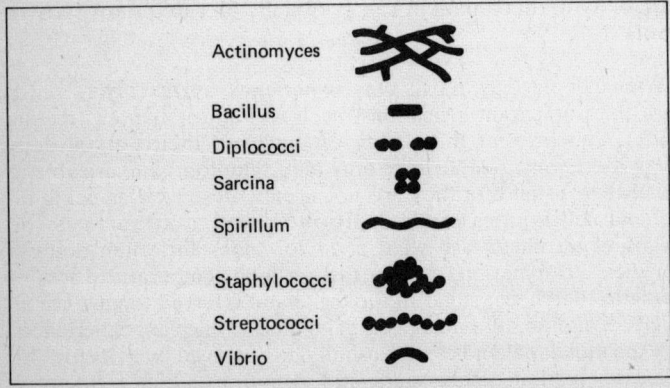

Figure 79. The shape of different kinds of bacteria

production of cheese, yoghurt and beer and in the treatment of sewage. Examples of bacteria that are pathogenic to man are the salmonella, staphylococcus, typhus and tubercule bacteria. Bacteria are usually grouped according to shape (see Figure 79).

2. Viruses. Regarded by some as non-living in that they do not have a cellular structure but are combinations of just protein and nucleic acid, these are the smallest type of pathogen and live inside the cells they have invaded. The common cold is caused by a virus. Other viral diseases are smallpox and poliomyelitis (see page 67).

3. Protozoa. These are also unicellular. They are usually found in water. Very few of them cause disease. An example of a pathogenic species is the amoeba which can cause amoebic dysentery.

4. Fungi. Most of these are harmless. A few do, however, give rise to infections, mainly of the skin. One of the most common fungal infections is tinea pedis, better known as Athlete's foot (see page 183).

NOTE: These organisms should not be referred to as 'germs' in any examination paper.

The body is protected from infection in the following ways:

1. **The skin** secretes a protective (antibacterial) oil called **sebum** and acts as a physical barrier to invading pathogens (see page 103).

2. Tissues or glands associated with the body's openings, or orifices, such as the mouth, eyes, ears and genitals, resist the entry of harmful organisms by secreting a fluid which acts as a sticky barrier covering the cells exposed to the external environment. Micro-organisms find it difficult to penetrate this **protective fluid** and so are unable to gain access to the cellular membranes or the blood vessels beneath. For example, the lachrymal glands of the eyes secrete on to the cornea lachrymal fluid, or tears; the outer canal of each ear is coated with wax from the ceruminous glands; the salivary glands in the mouth secrete saliva; the stomach and gut lining secrete digestive mucus, and the lining of the vagina and the rectum each secrete a similar fluid. These secretions are also **antiseptic** (see later).

3. Any micro-organism which does gain entry to the blood through a cut or a wound is likely to be attacked and ingested by **phagocytes** (see page 74). In addition, the lymphocytes produce protective substances called **antibodies** which are carried to all parts of the body in the circulation. Following an infection, antibodies are produced in response to a pathogen (or any foreign substance), known as an **antigen,** and can remain in the blood for months or even a lifetime. They will provide immunity against the effects of a re-infection by the same antigen (see page 78).

Immunity can thus be acquired naturally. But it can also be acquired artificially by a process of **vaccination.** Vaccines are small dilute doses of dead or weakened microbes or their toxins. Introducing a vaccine into a healthy person's body orally or by injection stimulates the formation of specific antibodies as if the person had been infected normally.

Because most people remain in good health for most of their lives the body's immune system is obviously very good at recognizing invaders and protecting us from them. Unfortunately, perhaps, this system is so efficient as to recognize normal cells or substances of another person's body as 'foreign', attacking them too. In the past, this caused problems when organs such as the heart or kidneys were surgically transplanted from one person to another. The immune system of the person receiving the transplant (the recipient) would perceive the new organ as an unwanted intruder and reject and attack it. Surgeons have more or less solved this problem by taking care to transplant organs only into people who have a similar type of tissue (possess the same type of natural antigens on the surface of their body cells) and by administering special drugs called **immunosuppressives** which reduce the activity of the lymphocytes.

Allergy is the term used to describe the immune response when antibodies formed against an antigen erroneously cause damage to the body's own cells. Hay fever is a common allergy the symptoms of which are sneezing and a runny nose. Some people are allergic to certain proteins such as eggs or shellfish and may develop a rash or vomit each time they are exposed to their particular 'foreign' foodstuff. Antigens which cause allergies are called **allergens.**

How diseases are spread

There are six main routes by which infection is spread:

1. By airborne droplets. Water droplets from the respiratory passages of individuals are projected several metres during exhalation, talking, sneezing and coughing. In the case of infected people, the droplets are likely to contain micro-organisms. Infected droplets may dry out, leaving the micro-organisms capable of infecting another person when rehydrated later.

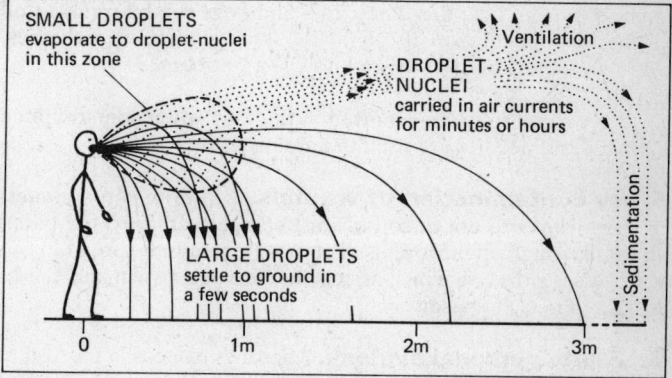

Figure 80. The spread of infection by airborne droplets

Spores of bacteria and fungi may also be present in ordinary dust. If dust containing pathogens is blown about and breathed in by a susceptible individual, infection will follow.

2. By ingestion. Infected matter may be swallowed. Food which is contaminated by micro-organisms can cause food-poisoning (see later).

3. By direct invasion. Skin diseases can occur when pathogens directly penetrate the surface of the skin. Moist skin is particularly vulnerable, and infections can be picked up from, for example, swimming baths which are insufficiently chlorinated or from walking barefoot on wet contaminated floors. Athlete's foot, a fungal infection which causes redness, itching and fissures between the toes, can be caught and spread rapidly from person to person in this way.

The disease known as schistosomiasis, which is endemic in Africa, is picked up by swimming in water that is infected by snails carrying the parasite *Schistosoma* (also known as the blood fluke).

Sexual intercourse is the means by which venereal (sexually transmitted) diseases are spread. The commonest ones are now gonorrhoea, syphilis, trichomonas vaginalis, chancroid and herpes genetalis. Hepatitis, an infection affecting the liver, can also be spread by sexual activity (see also page 190).

4. By inoculation. Pathogens carried by infected animals such as mosquitoes, fleas, lice, rats and dogs can be transmitted to man by inoculation through the skin by bites or scratches. Animal carriers of disease are known as **vectors.**

Diseases can also be transmitted in this way when contaminated syringes and needles are used to administer injections.

5. By contamination of wounds. Airborne contaminants or organisms present in soil or dust can gain entry to the blood stream through open wounds. Tetanus (lockjaw) is spread in this way; this is a disease which affects the nervous system and causes sustained muscle spasms.

6. Faulty personal hygiene. Microbes can easily pass on to hands after defaecation, so hands must be thoroughly washed after going to the toilet. Faeces and other potentially contaminated waste from human habitations must be disposed of safely (see page 195).

Examples of transmissible diseases

The majority of transmissible diseases are spread by airborne droplets.

Diphtheria is an infection of the throat caused by a species of bacterium which is usually inhaled. The bacteria produce a powerful toxin that spreads in the bloodstream to other parts of the body. The infection can interfere with breathing and be fatal. In Britain, children are immunized against diphtheria, so the disease is now rare.

Rubella (German measles) is a viral disease spread by droplet infection. It is usually a trivial condition and needs no treatment. However, if it is caught by pregnant women during the first three months of pregnancy, their unborn babies can be affected by the virus such that they are born with a disability, most commonly deafness. All girls should therefore be immunized against rubella before they are old enough to have babies.

Tuberculosis (TB) is caused by the tubercle bacillus. At one time this disease was the most common cause of death. Now most victims recover after treatment with antibiotics (see page 190). Tuberculosis can be spread by airborne droplets or can be ingested through infected milk. All milk in Britain is now pasteurized (heated) so that any tubercle bacilli are killed. Special care has also been taken to rear tuberculosis-free dairy herds. The disease first affects the lungs but can attack other organs of the body and may affect the joints.

Typhoid is an infection of the digestive system caused by the *Salmonella typhi* bacterium. It is spread by water or by flies which have been in contact with contaminated faeces. It can also be spread by contact with hands infected with the bacteria. The disease is now confined to underdeveloped countries where the supply of drinking water is not purified.

Food poisoning. One of the most common communicable diseases in Britain is gastroenteritis, or food poisoning, which is caused by eating foodstuffs which are contaminated by various bacteria. The *Salmonella* group of bacteria is the biggest single causative agent of food poisoning. The main source of salmonellae is the digestive tract of animals, where these organisms live naturally aiding digestion. Salmonella from animal excreta may be transferred to food by man by direct handling, by touching cooking utensils with contaminated hands or indirectly by flies feeding indiscriminately on both food and the faeces. Another common organism implicated in food poisoning is the bacterium *Staphylococcus aureus*. These microbes often exist naturally in the nasal passages and skin folds of healthy individuals (without causing diseases) and so can rapidly invade the body through cuts, grazes and lesions of the skin. For this reason, cuts on the hands should always be covered with a waterproof dressing before food is prepared (see Food hygiene, page 189).

Malaria is caused by a protozoan parasite called *Plasmodium* that is spread from person to person by female mosquitoes of the genus *Anopheles*. The mosquito carries the protozoa in its salivary glands and injects them into the human bloodstream when it bites through skin in order to feed on blood. The disease causes a high fever with headaches and vomiting.

The parasites (the term used for organisms that depend on other living creatures for their survival) multiply in the liver and later on in the red blood cells. They cause occasional but regular fevers which over many years gradually weaken the body. The protozoa can be killed by anti-malarial drugs.

Prevention of malaria depends upon knowledge of the life-cycle of the parasite (see Figure 81).

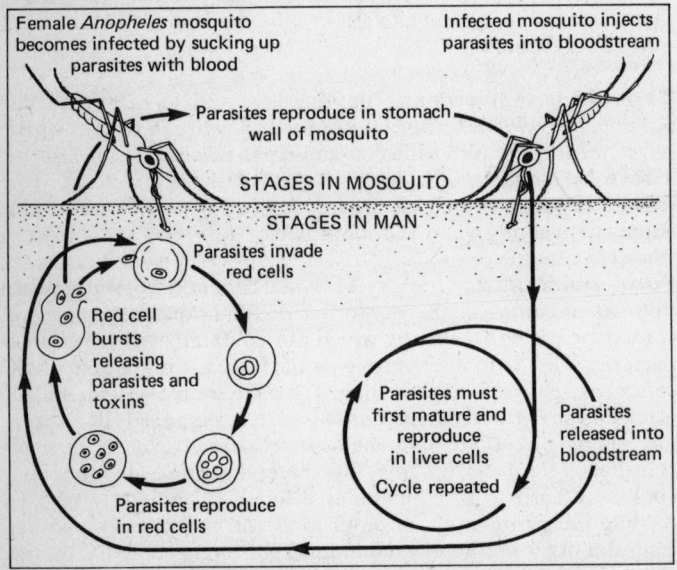

Figure 81. Life-cycle of the malarial parasite

The following measures can be used to control malaria:
1. Still water in lakes, ponds, swamps, roof gutters and drains should be removed or sprayed with a thin layer of oil containing insecticide. Because mosquitoes lay their eggs in still water, this measure suffocates the larvae which need to breathe at the surface of the water.
2. Fixing mosquito nets over doors and windows at night to prevent the adult feeding on man when asleep.

3. Spraying houses with insecticides to kill the adult mosquitoes.
4. Routine administration of anti-malarial drugs to kill any parasites that may have entered the bloodstream. Examples are chloroquine and proguanil which have superceded quinine, a drug used for many centuries against malaria.

It has not yet been possible to completely eradicate malaria. Drugs are expensive though effective, and some native people are unable to take them regularly enough. Various worldwide campaigns by the World Health Organization (WHO) have reduced the number of cases from about 250 million per year to about 100 million, using some of the methods listed above.

Typhus, a disease caused by a bacterium-like organism and that is marked by high fever and delirium, and other similar diseases are spread by lice. Hepatitis, or inflammation of the liver due to a virus or protozoan infection, can be spread through the use of unsterile syringes and ear piercing equipment (serum hepatitis) or contaminated food and drink (infectious hepatitis).

The sources of transmissible diseases

An epidemic is a sudden unexpected rise in the number of cases of a disease occurring in a particular place. When an epidemic disease spreads to several countries and the number of reported cases rises to very high levels in each of these it is said to be **pandemic.** Some diseases are referred to as **endemic.** This means they occur normally in a particular country and a steady number of cases is reported each year. For example, influenza is an endemic disease in the UK.

Human sources of infection are mostly patients who are in the **incubation period** of the disease (i.e. the early stage when there are no signs or symptoms of illness) or who are actively suffering from it. **Carriers** of a disease are either healthy people who harbour the pathogen without ever feeling ill or they are patients who have recovered from an infection but who go on harbouring the pathogen and pass it on to others.

Animal sources are those creatures which carry human pathogens, usually without being ill themselves. For example, fleas can transmit plague, flies can transmit food-poisoning, and body lice, fish, chickens, mosquitoes, parrots and cats are all implicated as carriers of pathogens, or vectors. Details of the life-cycle of some of these organisms are contained in Chapter 10 for those candidates whose syllabuses require them.

Soil contains the pathogens causing tetanus, gangrene and botulism, three very serious diseases which can be fatal.

Prevention and control of transmissible diseases

The health of a community is usually judged by the frequency with which diseases occur. Information is collected about people as they are born, marry and die so that the size of the population is always known. Records are also kept of conditions treated in hospitals, of occupational diseases and of accidents.

All cases of highly communicable, or infectious, diseases are required by law to be notified to the DHSS, the local medical officer for environmental health or to the public health laboratory service. Immediate action can then be taken to prevent the spread of disease. This can involve:

1. Identification. The source of infection or carrier of the disease must be identified and destroyed or treated. It is important that known carriers should not be allowed to take up occupations where they may infect others. For example, they should be prevented from preparing or serving foods and administering injections.

2. Isolating infected individuals. In certain conditions it is sufficient if the individuals avoid close contact with others until they are no longer infectious. In highly infectious cases, patients are **'barrier'** nursed in a room (an isolation room) away from other people. When inside the patient's room nurses wear special protective clothing and take care to see that no pathogens can be taken outside of the room.

Anyone with a venereal disease must not have sexual intercourse until he or she is clear of infection.

3. Quarantine. The word quarantine is derived from the French word for forty (quarante) and originally meant isolation for 40 days. Anyone who has been in contact with a person suffering from a highly infectious disease must be isolated for the length of time known to be the incubation period for that disease. This is a precaution against that person developing the disease and becoming infectious. Sometimes treatment of the person is begun before possible symptoms develop – this is the case with someone who has had sexual intercourse with an individual known to have a venereal disease.

4. Immunization. A mass vaccination programme of susceptible individuals may be implemented to prevent the disease spreading.

Extra precautions. Wherever an infection is present, extra trouble should be taken to see that pathogens do not flourish. Surfaces should be sterilized by washing with chemicals that kill bacteria (disinfectants); clothing or bedding may need to be burned and should at least be boiled or autoclaved (a process using steam heated to a high temperature). Good personal hygiene is essential. Towels should not be shared. The skin should be dried thoroughly. Hands should be washed after touching anything which could be contaminated. An antiseptic will help to kill microbes but if one is used, care must be taken that it does not damage human skin.

Food Hygiene. Bacteria are not usually dangerous unless they are present in large numbers. Unfortunately, they tend to reproduce very quickly, some of them doubling their numbers every 20 to 30 minutes. They prefer warm, moist conditions, not too salty, sugary or acidic. Meat, poultry, pies, stews, gravy, egg dishes and dairy products in normal conditions all favour the growth and reproduction of bacteria. To ensure that foods do not become contaminated by micro-organisms present in air, it is important that they are consumed quickly. Also, provided their taste or consistency is not affected, foods should be kept in a refrigerator wherever possible, and raw food should be separated from cooked foods to prevent bacteria passing from the former to the latter. Ideally, cooked food should be covered, especially outside the fridge, to prevent access to flies.

Some foodstuffs can be preserved longer than normal if they are deep frozen, dried, smoked or treated by the addition of chemicals. To prevent bacteria being passed from hands to food, the hands should always be washed after handling raw meat and vegetables and after every visit to the toilet. This rule is of obvious importance to food handlers. Other precautions are not using kitchen utensils for more than one meal without washing them in between; washing up dishes and utensils in very hot soapy water and then rinsing them; and keeping the kitchen and its work surfaces spotlessly clean. Waste matter should be put in plastic sacks then into a container with a lid to prevent the entry of flies.

Medical treatment of transmissible diseases
Serious infections such as rabies can be treated by injection of

preformed antibodies against the invading pathogen. This is done because the body takes time – at least five or six days – to produce its own antibodies. During this time the patient can die of the infection. The more usual treatment is to prescribe a course of **antibiotics.** These natural substances inhibit the growth of bacteria, thereby successfully reducing an infection. Most antibiotics are produced by certain species of moulds (fungi) as a weapon against bacteria in the competition for food in the soil. An example is **penicillin,** which Alexander Fleming discovered in 1928 to be produced from the mould *Penicillium notatum*. Other antibiotics are produced by one bacterium against other competing ones, for example, **streptomycin,** which is produced by members of the Streptomyces genus of bacteria. There are over 50 different antibiotics, most of which are now manufactured synthetically from their chemical components. Many new ones are currently being developed (as part of the science of biotechnology).

Antibiotics can be classified according to the range of their antibacterial action. Some have a **broad spectrum** of action. In other words, they can be successfully used against a wide range of different bacteria. Other antibiotics have a much narrower range of action, which can be particularly advantageous if there is a danger that the normal bacterial population of the body may be affected.

For a long time it was thought that antibiotics would eventually become useless because most bacteria tend to develop a resistance to their efforts. This has not become too serious a problem because so many new antibiotics have been developed to replace the original ones. Today it is even common medical practice to give antibiotics to patients to prevent infections developing, as with some venereal diseases (see page 184). The use of a drug in this way (to prevent rather than cure infections) is known as **prophylaxis.**

Antibiotics have no effect at all on viral infections, so you should not expect a doctor to give you them every time you have an infection. You should never take antibiotics because "you had a few left in a jar". They are powerful drugs which should only be taken on medical advice. It is also important that you complete the whole course of an antibiotic; you will be instructed to continue taking the drug even after the symptoms have subsided. This ensures that the infection does not re-start.

Some patients develop an allergic reaction to antibiotics. Other side-effects result from taking too high a dose of the drug. These include deafness and kidney damage.

Largely through the success of the antibiotics there are now very few adult deaths in the UK simply from transmissible diseases. Today's killers are the diseases of the blood circulation and cancer, both non-transmissible diseases.

Non-transmissible diseases

Cancer. This is any growth of abnormal tissue that arises from the disordered and uncontrolled division of cells that invade and destroy the surrounding tissues. These cells are referred to as **malignant.** Cancers arise for reasons which are not yet fully understood. If the disease is not checked quickly, malignant cells spread to other organs and sooner or later cause death. Many cancers can be detected at an early stage, however, and treated very successfully by surgery and by drugs.

The most common form of cancer in men is that of the lungs. It usually causes death within 5 years of diagnosis. Lung cancer can be caused by cigarette smoking. Although some people with lung cancer have never smoked, nine out of every ten people who suffer from it are smokers. The most common form of cancer in women at present affects the breasts. If a cancerous lump is discovered in the breast at an early stage, there is a good chance that it can be removed before the cancer spreads.

Until now, fewer women than men have become cigarette smokers. However, today there is a trend for less men but more women to take up smoking, so that by the year 2000 lung cancer will probably overtake breast cancer as the most common cause of cancer deaths in women.

Circulatory diseases. A combination of thickened blood, narrowed arteries and fragile blood vessels can lead to heart disease, high blood pressure and strokes. As people get older their blood vessels become less elastic. If the vessels are too constricted and blood is too thick for some reason, then blood pressure is raised. Narrow arteries cause the blood to flow more slowly and enable a clot, or **thrombus,** to form. This may later detach and block, say, the coronary artery in the heart or an artery in the brain; the thrombus is then referred to as an **embolism.** A thrombus in an artery to the brain can lead to a **stroke,** and in an artery to the heart (a coronary thrombosis), to a **heart attack.**

Smoking, poor diet and lack of exercise all play a part in contributing to circulatory diseases. Chemical substances and carbon monoxide in cigarettes have a damaging effect on heart muscle. Smoking also increases the amount of fibrin in blood, making it thicker and more likely to form dangerous clots. Diets which contain too much saturated fat and too little fibre (roughage) tend to produce **atheroma** (plaques). This is a substance which becomes laid down in arteries making them narrower in diameter. Lack of exercise results in a slowing of the circulation and causes blood to be thicker than is desirable.

Accidental deaths. For children in Britain over one year of age, accidents are the main cause of death. Up to the age of five, home accidents are the commonest cause of death. Older children run additional risks of being killed during play outside the home or from involvement in road traffic or farm accidents. A common fatal accident in boys over the age of sixteen involves the riding of motor cycles. Seventy per cent of home accidents which receive treatment in hospital accident and emergency departments are from cuts, bruises, poisonings, burns and scalds.

Parents and older brothers and sisters are responsible for the safety of the under-fives at home. Everything in the home must be considered as to whether it could cause harm to a young child. If it could, it should be removed out of the child's reach. An unguarded fire, an iron left switched on, a bottle of household bleach, a bottle of aspirin tablets, an open window, a pair of scissors, a bread knife, a polythene bag, a slippery floor, a pan of boiling soup on the edge of the stove; all these can cause accidents to children. As soon as they are old enough, children should be taught how to cross roads safely and not to touch dangerous machinery such as is found on farms. Medicines and sharp instruments should always be kept out of reach. Anyone who wants to ride a motor cycle would be well advised to look at the statistics of young people who have died whilst riding them. Many of those who survive motorcycle accidents have been left severely disabled. It is important to take a course of instruction in how to ride cycles as safely as possible.

Mental illness. This includes all disorders of the emotions. Like physical illness, it can be acute or chronic. Symptoms vary enormously, but include wildly fluctuating moods, severe anxiety, headaches, prolonged unhappiness and sleeplessness. In more severe cases, behaviour is irrational and bizarre. Delusions occur, in which patients believe themselves to be persecuted by imaginary people. Other patients experience **hallucinations,** seeing things or hearing voices that do not in reality exist (although in the patient's world they are very real).

Neurosis is used to describe less severe acute mental illness such as anxiety states, phobias and depression. **Psychosis** is a more permanent state, characterized by neurotic symptoms and the more severe symptoms described above. This is a gross simplification of what is a highly complex, little understood area of ill-health. Mental illness, however, can be distinguished from mental handicap or mental disability. The latter has an organic physical cause which permanently impairs the functioning of the brain. Mental illness, on the other hand, can have an organic cause, but it is generally thought to be caused by family or environmental pressures which continuously apply stress to a person's life. Most mental illness is treated by drugs, although this only reduces the symptoms of a mental illness. A cure is only possible by treating the cause of the illness.

First aid
The quality and quantity of first aid given to an injured person, or casualty, can often be life-saving. This involves:
1. Ensuring there is a clear passage for air to ventilate the patient's lungs.
2. Stopping excessive loss of blood.
3. Carrying out artificial respiration if breathing has stopped (see page 68).
4. Covering the patient to maintain body temperature.

Obstruction of breathing is especially hazardous in the unconscious patient. Such patients require constant supervision and should be placed in a **three-quarter prone position** (see Figure 82) after vomit or blood has been sucked out of the lungs and mouth.

The chest can be supported by a pillow or blanket, but care must be taken to avoid pressure on the chest wall. The head is tilted up to provide a free passage of air to the lungs. The uppermost arm is flexed in front of the trunk, with the hand under the jaw to provide extra support. To prevent the patient from rolling over, one of the lower limbs is flexed at the knee and hip.

Virtually all wounds will bleed to some extent, either internally or externally, but major, life-endangering haemorrhage will usually

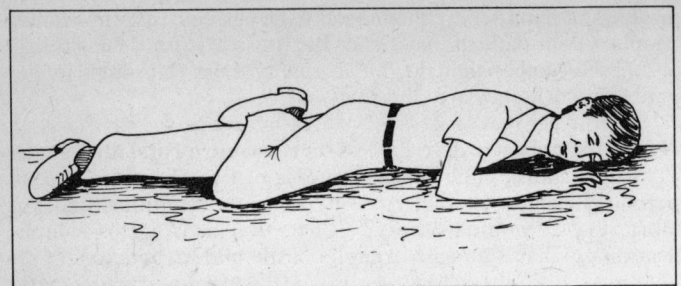

Figure 82. The three-quarter prone position

only result from laceration of a major blood vessel. Severe internal bleeding can only be stopped by surgery, so removal to a hospital is urgently needed. External bleeding will usually stop after firm application of a dressing over the wound for 5 minutes. But bleeding from major blood vessels demands extra measures. Direct pressure on the damaged vessel within the wound can be applied by pressing hard on to a pad or dressing. The pad should then be tied firmly in position. Other methods such as applying a **tourniquet** should not be used until it is clear that all else has failed to stop loss of blood. A tourniquet stops the flow of blood by compressing the artery against a bone. If left on for any length of time it will stop all blood supply to a limb (which will then later have to be amputated), so a minute or two is about the maximum.

If any bones are fractured the patient should not be moved until the fracture has been supported by a splint. Burn wounds should not be touched and clothing not removed unless it is smouldering or soaked in chemicals.

Health of the community

The general health of a population depends on many factors but most importantly:

— the physical features and climate of the environment in which the people live.
— the standard of housing and sanitation.
— the range of foods available and general level of nutrition.
— local industries and any pollution caused by them to the air and water supplies.
— the level of immunity to diseases among the people.

— the provision of health and social services and the awareness of how to use them.
— knowledge of how to lead a healthy lifestyle.

1. Physical features and climate of the environment.
These factors govern the natural flora and fauna and hence parasites which exist in a community. They also effect the outdoor activities likely to be carried out by populations, the clothing they wear, the housing they live in and the nature of local industry. Geographical features influence the ease with which one population can visit another and consequently how quickly infectious agents can be passed between groups.

2. Standards of housing and sanitation. The purpose of housing is essentially to provide shelter. Its design and construction will depend on the local climate and availability of building materials. In Britain, housing is constructed mainly of bricks, mortar and concrete. Roofs are sloped steeply and made from overlapping tiles to keep out rain. In order to stop rain penetrating the sides of the house, the outer wall is made double; the cavity between the two walls prevents moisture crossing to the interior (and provides insulation). To stop moisture rising up the walls from the ground, a **damp course** made from a layer of impermeable material is incorporated into the lower part of the walls. Inside the house, the floorboards are raised above ground level so that air is permitted to circulate in the cavity under the floor. Air enters the cavity through **ventilation bricks.**

Most houses in Britain have running water and flush toilets for the disposal of human excrement. Water is obtained by local authorities from pure sources. It is then filtered and purified further by the addition of chlorine before it enters the domestic water supply (see also page 202). Water which is suitable for drinking straight from the tap is called **potable** water.

Sewage (waste water) disposal begins when the domestic toilet is flushed. About 15 litres of water are released each time you 'pull the chain' (or nowadays press the cistern handle), sweeping faeces and urine into the main drains. The U-bend in your toilet outlet pipe traps a small quantity of water which acts as a seal preventing bad smells getting back from the main drain into the house. Sewage is carried away from houses in underground pipes, or **sewers,** where it becomes mixed with waste water from factories, offices etc. Storm, or rain, water also passes into sewers through street drains and eventually all sewage arrives at the local sewage treatment works (see page 200), sometimes having to be pumped there if the sewers do not run downhill.

Rubbish disposal is carried out by local authorities too. Each household discards its waste into a dustbin which should be covered to prevent dogs, cats, rodents and flies gaining access to its contents. Rubbish is collected from houses and taken to the local authority tip where it is separated into organic and inorganic matter. The former is converted into compost or fertilizer; the latter is pulverized and compressed before being buried. Organisms in the soil act on buried rubbish and bring about its decomposition.

3. Food and nutrition. The range of foodstuffs available to a population will affect the general state of health. People who do not eat enough of the right kinds of foods become more susceptible to disease (see Chapter 2).

Some terrains do not support the growth of plant foodstuffs and so these have to be imported. Malnutrition is more likely among poor people who cannot afford to purchase the necessary food to ensure health or who are not sufficiently educated to know what sort of diet is necessary for health.

4. Local industries and environmental pollution. The type of industry found in a community often depends on deposits of minerals in the area or crops or animals which thrive in that locality. For example, coal deposits in Wales and the North of England led to the development of the coal mining industry in those parts; flax grows well in Ireland and supports the linen industry; and in Scotland sheep grow long woolly coats and support the woollen trade there.

Industries release by-products, usually in the form of smoke, dust or chemicals, which may pollute the land, water or air. For example, sulphur dioxide is released when coal or oil is burned. When mixed with rain it forms sulphurous acid which causes the stonework of buildings to decay and which makes soil, lakes and rivers more acid. Crop cultivation is damaged and the growth of freshwater fish is stunted. Factories have very tall chimneys to ensure that polluted air is carried away from the local community at a high level. Unfortunately, it can mean that its effects are felt in a neighbouring community.

The clean air legislation passed in Britain in the 1950s created a number of smokeless zones in industrial areas where only smokeless fuel could be burned. This did a great deal to control air pollution and prevented many people suffering unnecessarily from the respiratory illness chronic bronchitis which was so widespread in the North of England that it used to be known as 'the English disease'.

Air pollution by factories is also controlled to some extent by factories washing their fumes in a strong current of water to remove the sulphur dioxide before releasing them into the atmosphere. Other forms of industrial pollution stem from the release of metal wastes which enter the soil, rivers or sea where they may contaminate fish.

5. Level of population, or 'herd', immunity. The resistance of a community of people to the spread of infection is called herd immunity and it depends on the proportion of people in a group who have acquired immunity either by natural or artificial means to the infective agent. Obviously, the more people who are immune, the less people there will be available to contract an infection and pass it on.

The degree to which infections may spread depends on the frequency with which new infections are introduced; the opportunities afforded for infected people to mix with susceptible people; the ease with which the particular infection can be passed on; and the length of time that one person remains infectious.

Epidemics occur when the level of herd immunity to a particular infection is low. The number of cases of people who contract the infection is then much greater than usual.

6. Provision of health and social services. The prevention and treatment of disease depends upon the extent of health and social service provision.

The National Health Service (NHS). In Britain, comprehensive medical care has been available to all residents, free of charge, since the formation of the National Health Service in 1948. The Health Service is paid for mainly by general taxation; less than 10 per cent of its costs are covered by the National Insurance contributions paid by employees and employers. About 3 per cent of costs are covered by charges made for prescriptions, dental, ophthalmic and chiropody services. No charge at all is made to some people such as old age pensioners and unemployed people receiving supplementary benefit.

General practitioners (GPs or family doctors) and their nurses (health visitors and practice nurses), dentists and pharmacists are contracted to work in the community providing primary health care, i.e. they treat patients when they first become ill or need attention, or when they can be cared for at home. The GP arranges for specialists' opinions and for **hospital treatment** either as an in-patient or out-patient. The NHS provides all hospital and specialist treatment needed for sick people. It also provides services to prevent people from becoming ill (prophylaxis). For example, it keeps a check on pregnant women in **antenatal clinics** to make sure mothers and babies will not experience difficulties when the babies are actually delivered. It also provides a routine vaccination and check-up service for all young infants. **Child health centres** then continue to monitor the development of young children.

All the above services are administered by the Department of Health and Social Security (DHSS). The administration of the NHS was reorganized in 1974 and modified again in 1982. The organization is now managed by three levels, as shown in Figure 83. Various committees are involved whose members meet to report on the running of services and to plan for improvements.

Local Authority services. In addition to the health services provided by the NHS, some facilities are administered by local authorities and paid for out of the rates charged to local householders. These services are mainly for the elderly and the physically or mentally disabled who need care to enable them to live in the community but do not require the attention of hospital doctors or nurses. Residential units are provided for those who cannot look after themselves. Day centres, meals services ('meals on wheels'), domestic helps and mechanical aids such as wheelchairs are provided for people who can live at home.

7. Knowledge of how to lead a healthy life. There is no doubt that people who know how to look after themselves properly are healthier than those who do not. It may seem that common sense is all that is necessary but this is not the case. Nowadays, District Health Authorities employ **Health Education Officers** whose job it is to raise the general level of knowledge about health.

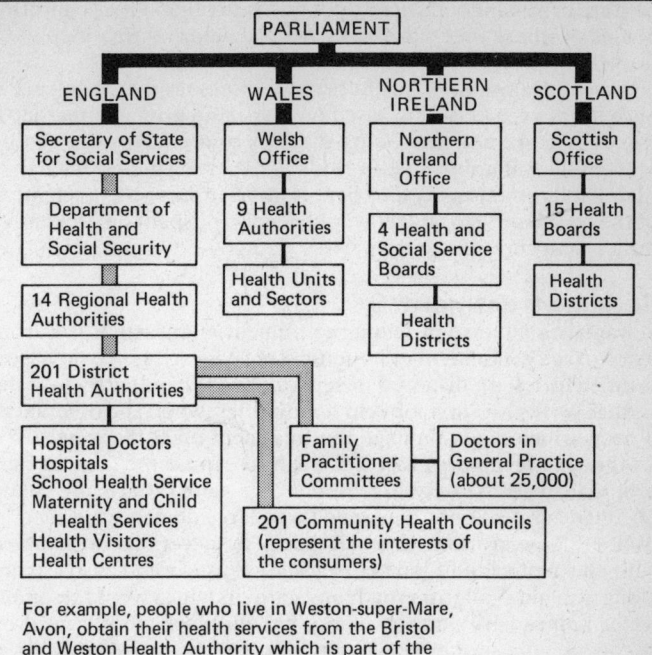

Figure 83. Organization of the NHS

This is done by advertising campaigns, lectures, posters and films and the work is often concentrated in the schools where more and more pupils study health education.

In 1968, the British Government set up the **Health Education Council** (HEC) to promote knowledge about health among the general population. This is done primarily through television and other means of advertising. The HEC produces many posters and leaflets which are free of charge. It is particularly concerned with informing people about the risks they run if they smoke cigarettes, drink too much alcohol, overeat and take too little exercise.

The HEC encourages people to look after their teeth, to eat a sensible diet and to have the vaccinations which are necessary to increase the population immunity to common diseases. It also tries to prevent the spread of diseases by requesting that people who are ill stay indoors and do not mix with others. It also informs people of where they can obtain medical help and treatment.

Health education attempts to prevent disease rather than cure it once it has occurred. It is based on the sound principle that most UK deaths are now caused by diseases or accidents that can be prevented. But unfortunately this means getting people to change their habits and the way they live. If this is to succeed, much more of the health service budget would need to be spent on preventive rather than curative services.

Sewage treatment

Sewage is a mixture of human excrement and waste water from offices, factories and other buildings (see page 195). (Toxic wastes from factories are disposed of separately.) When it arrives at the sewage works, the first job is to remove rags, wood and other large objects which might damage the treatment plant machinery. To do this, the sewage is passed through screens in the form of bars of metal spaced closely together. Large solid objects are raked off and destroyed or burned, and those large objects which can be broken down are torn to small pieces in macerators and mixed with the more liquid part of sewage. Waste water also carries along grit and sand from roads and gardens which would wear out water pumps very quickly. These fine particles are removed by conducting the sewage slowly through channels so that they drop to the bottom of the channels and can be dredged off. They are then washed and used to fill large holes in the ground left by excavation work.

The next step is to remove as much of the semi-solid matter (e.g. faeces) as possible. The sewage flows into large **sedimentation tanks** where this material (called **crude,** or **raw sludge**) settles at the bottom. The sludge is swept by electrically driven scrapers into a hopper and then pumped to the **sludge 'digestion' plant.** The remaining liquid, or **primary effluent,** is treated in a secondary treatment unit or **aeration tank.**

In the aeration tank microbes are allowed to grow which feed on the waste matter in the effluent and destroy it, leaving only gases and water. They need a lot of oxygen from the air to do this, so compressed air is passed into the effluent through special tiles (called diffusers) at the base of the tank which have very small holes so that the air enters as small bubbles. It takes about 8 hours for the microbes to destroy most of the impurities. The microbes

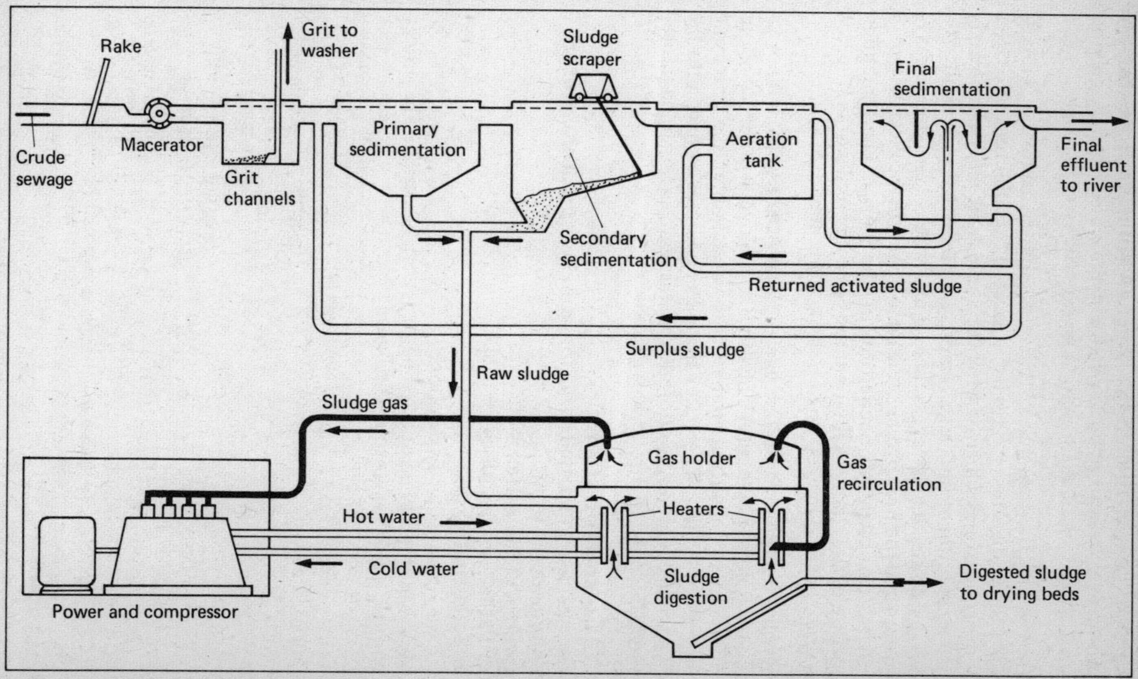

Figure 84. A modern sewage works

are then separated from the water in **final sedimentation tanks** so that they can be re-used, and the water which is left (the final effluent) is clean enough to go into the river or sea.

Meanwhile, the raw sludge that was pumped from the primary tanks to the digestion tank is being tackled by a different collection of microbes, ones which destroy the unpleasant 'smelly' materials and change them into a gas (mostly methane) that is similar to North Sea gas. This process takes 3 to 4 weeks. The sludge gas is burned and the heat produced used to generate the electricity and other forms of power needed to run the works. The remaining portion of the sludge is then pumped to a disposal works where it is dried out and bagged up as a soil conditioner.

Sewage treatment is just one part of the **water cycle,** but nevertheless it is crucial to public health, the biological potential of our rivers and the health of animals and plants both in the water and that drink from it. The water that has been passed back into the rivers will eventually evaporate and fall again as rain to be treated and used in the domestic water supply.

Water purification
To produce water that is fit for drinking (**potable water**) it should be free from:
1. Debris, soil particles etc.
2. Pathogenic microbes
3. Industrial wastes
4. Any noxious tastes or odours

Most water has to be purified before it can be drunk. The process of purification is shown in the figure below.

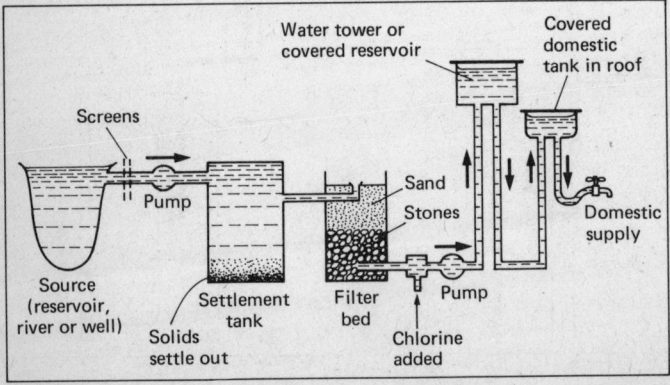

Figure 85. Water purification

A screen prevents entry of large objects, including fish. In the settlement tank the water is allowed to stand so that large particles can settle out. The water is then pumped to the surface of a settlement tank where it slowly filters through sand and stones which are covered with a film of microbes. These microbes break down organic material in the water; any that pass through the filter bed are killed by chlorine which is added to the water before it is stored in a covered reservoir.

Key terms

Acute illness Short-term, often life-threatening disease.
Allergen Something that stimulates an allergic response in the body.
Antibody Proteins produced by leucocytes in response to the presence of, and combatting, antigens.
Antibiotic Substance produced by moulds and bacteria that stops the growth of bacteria.
Antigen Substance that stimulates the production of antibodies; usually a protein.
Antiseptic A chemical applied to wounds to destroy or inhibit the growth of pathogenic bacteria.
Chronic illness Long-lasting disease, not usually fatal.
Diagnosis The method used by doctors to determine what is wrong with a patient.
Disability Permanent abnormal functioning of the body.
Epidemic Sudden outbreak of a transmissible disease that can spread rapidly throughout a community.
Immunity The ability of the body to resist re-infection by pathogens by production of antibodies.
Inoculation Injection of a substance into the body by piercing the skin.
Pathogens Organisms that produce disease or substances that are toxic to the body.
Prophylaxis Administration of drugs or vaccines to prevent rather than cure an illness.
Sterile Free of pathogens.
Symptoms The usually non-observable factors associated with being ill, e.g. headaches, nausea.
Transmissible disease A disease that can be passed from one person to another.
Vaccination The process of administering a vaccine.
Vaccine Preparation of dead, inactivated or relatively harmless pathogens that will stimulate the body to produce antibodies and thus protect against any future infection by that pathogen.

Chapter 10
Man and Other Living Organisms

This final chapter includes information which is not required by all examination boards. You should check your syllabus and study only those topics mentioned.

Ecology
This is the study of the relationship between living organisms and their environments. Within an environment there is a continual flow of energy and a cycling of matter. Energy enters the system in the form of light energy from the sun. This is converted by green plants into chemical energy within organic substances during **photosynthesis.** These substances are then used by all organisms (including the photosynthetic plants) in respiration to release the energy in a controlled way to drive the biochemical machinery of cells. The energy is eventually lost from the system in the form of heat.

Green plants, then, may be considered as **producers** of organic matter. Animals are **consumers** since they rely on the plants for their supply of organic materials, either directly, as with **herbivores** (plant eaters), or indirectly, as in the case of **carnivores** (meat eaters), which eat herbivores. In any habitat there are also **decomposers,** such as bacteria and fungi, which feed off the remains of plants and animals and in so doing release inorganic materials – the building bricks of organic biochemicals – back to the soil. Through these three methods of feeding the cycle of matter is achieved: producers converting inorganic material into organic matter and decomposers carrying out the reverse process. In any community, then, **food chains** are built up, the producers (green plants) being eaten by herbivore animals, which in turn are eaten by carnivores. For example, grass (producer) – cow (primary consumer) – man (secondary consumer).

Plants and photosynthesis. The process of photosynthesis is of vital importance to all living things since it is the only method by which organic materials are produced naturally. It has been estimated that an hectare of corn can produce 25,000 kg of sugar per year. During photosynthesis light energy is converted into chemical energy in the form of sugars. Sugars are stable compounds which

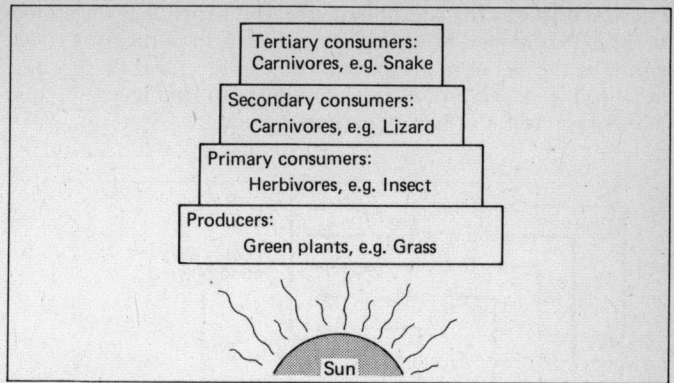

Figure 86. A simple food chain

all organisms can use as a source of energy. The significance of photosynthesis can be seen clearly in the carbon cycle (see page 206).

Photosynthesis can be represented by the equation:

$$6CO_2 + 6H_2O \xrightarrow[\text{chlorophyll}]{\text{light energy}} C_6H_{12}O_6 + 6O_2$$

(carbon dioxide + water → glucose + oxygen)

This equation simply indicates the raw materials needed for photosynthesis and the initial products of the process. Two points which it fails to make are, firstly, that it is not only sugars which are produced as a result of photosynthesis but also the carbon atom skeletons for carbohydrates, fats and proteins too, and secondly, that photosynthesis takes place in a large number of steps, each step requiring a specific enzyme. The details of the mechanism of photosynthesis were only established in the latter half of this century. This was a result of more sophisticated techniques being developed in other scientific fields, such as the use of radioactive isotopes.

Photosynthesis may take place in any part of the plant containing the green pigment **chlorophyll,** but in the majority of plants it is carried out mainly by the leaves as these are rich in cells packed with **chloroplasts,** the intracellular organelles that contain the pigment.

The carbon cycle. The amount of carbon dioxide in the atmosphere is maintained by a balance between the processes which withdraw the gas from air (photosynthesis etc.) and those which add it to the air (respiration and combustion). In Figure 87, solid arrows represent this flow of carbon dioxide.

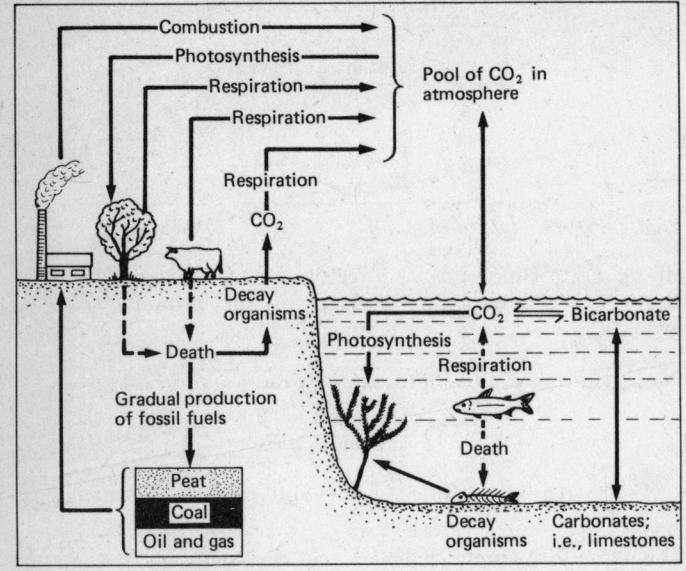

Figure 87. The nitrogen cycle

The nitrogen cycle. In poorly aerated soils de-nitrifying bacteria break down ammonium (NH_4^+) compounds resulting in the release of nitrogen (N_2) into the air. Nitrifying bacteria such as *Nictrobacter* and *Nitrosomonas*, may exist independently in the soil; these change the ammonium compounds into nitrates (NO_3^-). Nitrates are absorbed by plants. Nitrifying bacteria also exist in swellings, called **nodules,** on the roots of legumes (e.g. pea and clover). This is a **symbiotic relationship,** i.e. one in which both partners, plant and bacterium, benefit. The bacteria are able to use nitrogen from the air and build it directly into nitrogen compounds. Legumes are thus an important part of any crop rotation since the nodules they bear result in the plants increasing the nitrogen content, and hence fertility, of the soil.

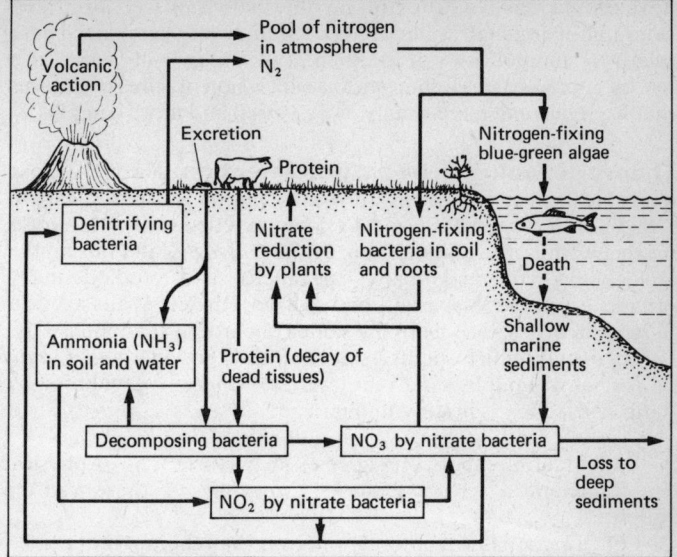

Figure 88. The antigen cycle

Man now influences considerably the balance of nature by interfering with these cycles. For example, the amount of combustion of carbon compounds has increased greatly in the last century, resulting in the rapid depletion of fuel reserves (e.g. oil, coal, wood) and an increase in the carbon dioxide content of the atmosphere. Man has also poisoned the environment with toxic wastes from factories and has replaced natural habitats with his own artificial environments, such as cities. This creates an imbalance which has already resulted in the extinction of many species and which threatens the existence of man himself. Through our knowledge of ecology we can achieve good and bad effects. It is to be hoped that in the future our improved understanding of natural systems will lead to a more rational and sensitive use of our environment.

Evolution

The theory of evolution explains how the enormous variety of plants and animals found in the world came into existence; that similarities between organisms are a result of descent from a common ancestor; and that differences between them are a result of variations accumulating between parents and offspring. It must be supposed that live structures capable of division were produced from non-living matter hundreds of millions of years ago. These relatively simple forms of life then, over millions of years, gave rise by a series of small changes to a succession of living organisms that have become increasingly more varied and more complex.

Theory of evolution by natural selection. The man whose name is most closely associated with evolution is Charles Darwin (1809-82). Between 1831 and 1835 he travelled round the world on the Admiralty Research ship *H.M.S. Beagle*, and during the voyage he collected many specimens and studied many organisms. In 1858 Darwin and Alfred Russell Wallace, who independently had come to the same conclusions, announced the theory of evolution by natural selection and Darwin's book *Origin of Species* was published a year later. The theory of evolution by natural selection is briefly summarized:

1. Individual members of a species show variations of physical and biochemical characteristics, and some of these can be inherited.
2. Offspring are always more numerous than their parents.
3. Despite this tendency to increase for any one species, the numbers tend to remain more or less constant.
4. As fewer organisms live to reproduce than are born, there must be a continuous struggle for existence.
5. Organisms which possess variations which are advantageous in a particular environment will be more fitted to live there. They will survive and reproduce and have offspring which may inherit the more successful variations. This is known as the **survival of the fittest.** There will be a tendency to diverge away from the original type by the accumulation of favourable variations.

Variation. The kind of variation which is important in evolution is that which the organism is able to hand on to its offspring and is transmitted through **genes.** This arises principally by **mutation;** by reassortment and recombination of chromosomes at meiosis; and as a result of fertilization. Man is the product of at least 1,000 million years of evolution and natural selection.

If you switch on the current to an electric light bulb the bulb will glow. Switch the current off and it stops glowing. A light switch and a bulb can therefore be on or off. There are no intermediate positions. The same principle applies to some human characteristics, for example the ABO blood group system. A person is either group A, B, AB, or O. He cannot be half-AB and half-O. When features, or characteristics, show a range of distinct differences like this they are said to be **discontinuous variables.** Another example is the inheritance of sex. People are either male or female. Such discontinuous variables are always genetically determined. They cannot be altered during the development of an individual. You cannot alter your ABO blood group by having a total blood transfusion. Your bone marrow will eventually replace the donor blood with blood of your own group which is determined genetically.

Most human features do not show discontinuous variation. Height, weight, skin colour, intelligence, athletic ability, for instance, are **continuous variables.** In any group of people there will be a wide range of intermediate types of any of these features. For example, if you give an Intelligence Quotient (IQ) test to a large number of people, most of them will score around the average 100 mark (see Figure 89). But some will score very low or very high marks. A distribution of this type is typical of characteristics showing continuous variation. Continuous variables are genetically determined to some extent, but are also strongly influenced by environmental factors such as diet, schooling, exercise, family life and housing.

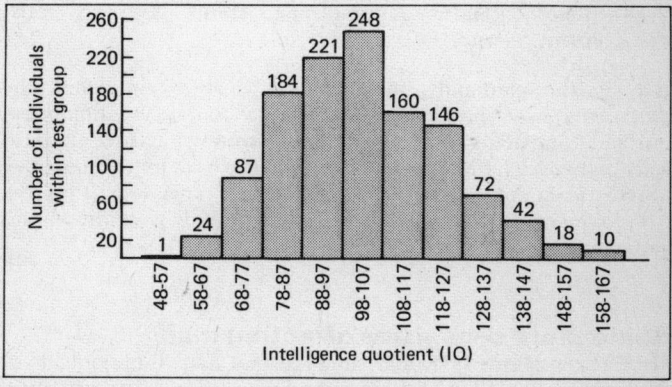

Figure 89. Continuous variation: Scores on an IQ test

Classification of the Animal Kingdom
Protozoa – Unicellular, e.g. *Amoeba, Paramecium.*
Coelenterata – Body made up of two layers of cells surrounding a central cavity whose only opening is the mouth, e.g. *Hydra*

Nematoda – Worms with unsegmented cylindrical bodies, e.g. pinworms
Platyhelminthes – Worms with small flattened unsegmented bodies, e.g. tapeworm
Annelida – Worms with a body cavity and two openings to gut, e.g. earthworm
Mollusca – Soft bodied with one or more shells, e.g. snails, mussels
Echinodermata – marine; shows radial symmetry, e.g. starfish
Arthropoda – Body segmented; many-jointed legs; exoskeleton of chitin, e.g. insects, crabs, spiders
Chordata (Vertebrates) – Brain in well-developed head; skeleton of bone or cartilage of which vertebral column forms central axis; tail
  Pisces (Fish) – Aquatic, move by tail and fins; gills, e.g. shark
  Amphibia – Partially terrestrial, eggs laid in water; moist skin; pentadactyl limb, e.g. frog, newt
  Reptilia – Terrestrial, eggs laid on land and protected by shell; scaly skins; pentadactyl limb, e.g. snakes, turtles
  Aves (Birds) – Feathers, forelimb adapted for flight; endothermic; eggs protected by shell; pentadactyl limb, e.g. gulls, penguins
  Mammalia – Hair; endothermic; young fed on milk, e.g. **man,** rabbit.

Among the mammals, man belongs to a group called the **primates,** which includes his nearest neighbours, the chimpanzee, gorilla, orang utang and gibbon. These **apes** (which include man) walk on two legs, which frees the hands to be used for manipulating objects and delicate precise, operations. This fact, plus the development of man's cerebral cortex, largely account for his success on the earth.

Some more organisms affecting man

1. External parasites are animals that live on the outside of the body and obtain their food from the host (i.e. man), usually in the form of blood. Two examples are the **flea** and the **louse,** both of which carry the very small bacteria-like organisms, the rickettsiae. Species of these cause the highly infectious disease **typhus.** Outbreaks of typhus can occur wherever people live in crowded housing with poor sanitation – 20 per cent of patients die unless treated with antibiotics.

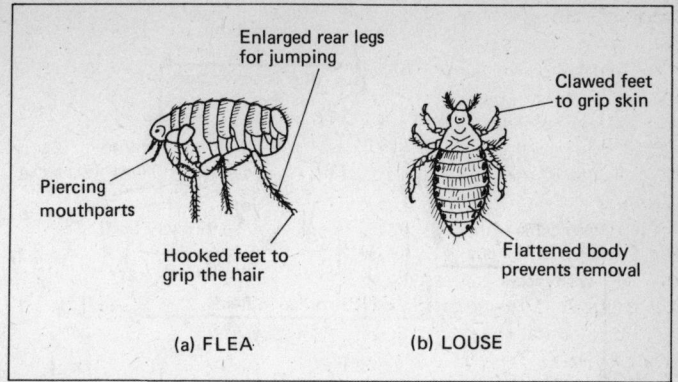

Figure 90. Human external parasites

The human flea and louse are both wingless insects with piercing mouthparts used to suck blood from the skin, which is how the rickettsiae are transmitted. There are three kinds of louse found in Britain:

The head louse lives in the hair and lays eggs called nits, which stick to the hair. They hatch out in about 6 days as nymphs, which are essentially miniature adults. After a further 10 days, the nymphs moult, become adult, and breed. Contrary to popular belief, the head louse is just as common in clean short hair as in dirty long hair, and spreads easily whenever people are in close contact (e.g. in schools). Treatment is by use of insecticidal shampoos and combing with a special comb to remove the nits.

The body louse is found only among those who rarely bath or change their clothes (e.g. tramps). It lays eggs on the clothes and is found on the skin only when feeding.

The pubic, or crab, louse is usually found in the pubic hair, and is often spread during sexual intercourse. Both the body and the pubic louse are otherwise similar to the head louse.

2. Internal parasites live inside the body. The **pork tapeworm,** *Taenia solium,* lives in the human intestine, attached to the gut wall by hooks and suckers on its head. It feeds off digested food and has no gut of its own. It reproduces by shedding mature segments packed with eggs which pass out with the faeces. If eaten by a pig, these eggs hatch out into larvae which bore through the pig's gut and form thin-walled sacs called bladderworms in its muscles. The bladderworms remain dormant until eaten by man, when they develop into the adult tapeworms (see Figure 91).

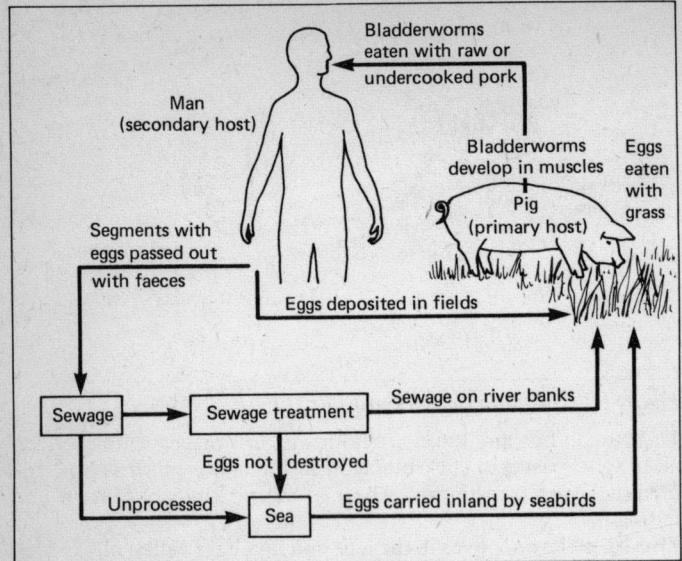

Figure 91. Life cycle of the pork tapeworm

The roundworm *Ascaris* lives in the ileum of man, feeding on digested food. The worm produces thousands of eggs daily, each with a resistant shell. If contaminated food or drink is ingested by another person, the eggs hatch into larvae which bore their way through the intestines into the blood vessels of the lungs. After further development, they wriggle back up the throat, where they are swallowed and then mature into the adult male and female worms in the intestine.

Pin worms (or thread worms) are small roundworms living in the human colon. They cause little harm except when the female moves out onto the skin around the anus to lay her eggs. Her movement causes great irritation, leading to itching. If the area is scratched, the eggs may become deposited under the fingernails and then swallowed so that reinfection follows. The eggs are sticky, increasing their adhesion to the skin. They are also blown about in dust and may be inhaled or eaten with food. The life cycle of the eggs on an adult human usually lasts about 5 weeks, so that fresh bouts of irritation occur every 5 weeks.

3. Viruses are sub-microscopic in size; they range between 0.0005 mm and 0.00001 mm. An isolated virus is a small speck that cannot grow or multiply except inside a living cell. Having invaded a cell, it takes over the cell's metabolism such that activity is diverted to the reproduction of the virus. Usually, the cell is destroyed and the new viruses released to infect neighbouring cells.

Figure 92 shows the life cycle of a **bacteriophage,** a virus that attacks bacterial cells and about which we have gained most of our knowledge of these essentially non-living organisms. A bacteriophage consists of a prism-like head containing a thread of DNA and a tail bearing several fibres.

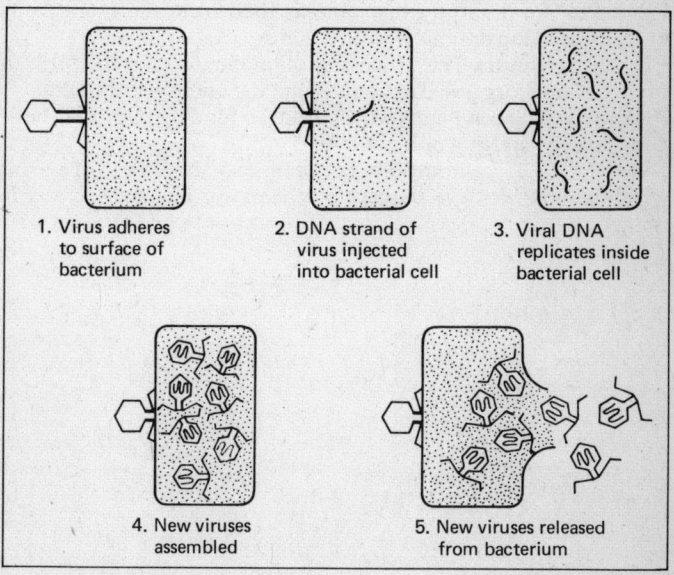

Figure 92. Life cycle of a bacteriophage

Some viruses produce toxins, but most damage cells just by upsetting cell metabolism. Some of the common viral diseases have already been mentioned. Viruses have also been shown to cause certain cancers. Antibiotics have no effect on viruses but the development of anti-viral drugs has shown great progress recently. **Interferon,** a natural anti-viral constituent of many secretions, is currently being researched and hopes are high that one form of it will soon be commercially available for a wide range of viral diseases. Acyclovir is a new drug with reported success in the treatment of herpes virus infections.

Key terms

Consumers In a food chain, organisms which live by eating others.

Decomposers Organisms such as bacteria and fungi which decompose organic matter.

Ecology The study of the inter-relationships between organisms and the way that energy is transferred between them.

Evolution The study of how new species of organisms gradually develop from other organisms.

Parasite An organism that obtains food from the living body of another living organism, called a host.

Photosynthesis The process of obtaining food in the form of sugars using inorganic compounds and the energy of the sun.

Producers Green plants which produce food by photosynthesis and form the starting point of a food chain.

Symbiosis A relationship between two different organisms which benefits both of them, e.g. vitamin-producing bacteria in the gut.

Examinations and Exam Technique

Having got this far you should know enough to pass your examination. But you still have to convince the examiners. Examiners are not allowed to mark scripts from schools they either teach in or are associated with in any way. So although they know your name, examiners will not know you, what you look like or how hard you have worked during your course. You have got only the marking time – about 6 to 10 minutes – to convince them that you are worth a pass, or better still, a good pass. All they have got to go on is what is in front of them, your exam script.

General advice

First impressions are important. Examiners appreciate answers that are neatly laid out and occupy the spaces provided. Handwriting should be clear and bold, so it is easy to read. If your handwriting is illegible, drill yourself to always write so that each letter is recognizable. A completely illegible script cannot be marked and will be returned. If you cannot help leaving ink blotches all over your paper, try using a biro instead.

Human Biology is one of the more difficult subjects to pass, the most common failing being imprecise or vague expression. Candidates often use commonplace explanations when more precise ones are needed. For instance, they write that vitamins are necessary in the diet because "they are good for you", rather than explain the diseases caused by vitamin deficiency. Or again, they use the word 'germ' when they mean pathogenic microorganism. Answers are often confused as well as imprecise. In answer to the question, "What hormone changes precede menstruation?" a candidate wrote that "The uterus has not been used so the body is getting rid of it, which occurs every 28 days".

While in all examinations correct spellings are very important, in science papers examiners tend to be a little more lenient providing the meaning is obvious. But no examiner would award a mark to the candidate who wrote that iron sulphate was added to drinking water "to give the body iron and stop amnesia". So be careful. (The word should of course be *anaemia.*)

Do not be tempted to cheat. Your ingenuity will almost certainly be matched by the examiner's. Penalties for proven cheating are severe. You can be sure that if you are found to be cheating on one paper you will be given a fail, and all of your other subject papers will be closely scrutinized.

All CSE and GCE O Level exams are marked objectively. This means that if a paper has a maximum mark of 60, each of those 60 marks will be allocated to specific points of fact (such as the name of an organ) or to specific reasoned explanations of fact (such as 'an enzyme is inactivated by high temperature because this changes the shape of the protein molecule'). No marks are awarded for irrelevant information, however lengthy.

If you missed the point or got the fact wrong you will not be awarded the mark. And what is more important, you cannot then recover that mark from another question. Each question will have a maximum number of marks. So if you fill your answer paper with the answer to one question only, you can get only the maximum number of marks for that question. Therefore candidates who do not complete the paper are at a grave disadvantage. Lets consider this point another way. If a paper has five questions, each carrying 20 marks, the maximum mark will be 100. Now if you answer only four questions, your maximum marked will be 80. Most candidates should expect to get half marks for each question, i.e. 50 per cent. But for the candidate attempting only four questions, 50 marks (the average pass mark) out of 80 is quite a tall order. It is always easier to get the first 50 per cent of the marks for each question. So always complete the paper. Work out beforehand how long you should spend on each question and stick to this schedule strictly. Check your schedule before you start writing to make sure you are answering the right number of questions. When under pressure, it is very easy indeed to misread the instructions at the top of an exam paper. Read the questions carefully before starting to answer them, and always read your paper through before you give it in at the end of the exam.

The written paper
Today, more emphasis is placed on the understanding of basic biological principles than was the case a few years ago. With many examining boards there has been a tendency to move away from the straightforward factual questions to those of a more searching nature. The following paragraphs should help you be aware of these and other types of questions.

Short-answer questions. A series of these may be set as a separate examination or they may form the first part of an examination paper. They can take the form of one-word, one-sentence, or one-paragraph answers. A definite amount of space is left in which to answer the question, indicating the length of answer that is required, i.e. a one-word answer is inadequate if four lines of space are given. Care must still be taken to ensure that the question is interpreted correctly. A typical short question could be, 'Give two differences in structure between arteries and veins'. An answer such as 'Arteries carry oxygenated blood and veins carry deoxygenated blood' will inevitably be given. These are not differences in *structure*. Examples of correct answers are: 'Artery walls contain much elastic tissue but vein walls contain very little', or 'Arteries do not possess valves, veins do possess valves'. In this type of question it is also important to refer to both the artery and the vein: for example, the answer 'Veins have valves' or 'Veins have little elastic tissue' would be unacceptable because no reference to the equivalent structure in arteries has been given.

Multiple-choice questions are part of some examination papers. Here the student is asked to tick, circle or underline the correct answer from a choice of four or five.

Example: Urea is made in:

     a) the pancreas; b) the bladder; c) the kidney; d) the liver.

Do not indicate two right answers or the question will automatically be marked wrong. If you are not certain which is correct, choose only one answer; at least there will be a good chance of getting it right and being awarded a mark.

Tabulate: means a table consisting of two or more columns should be drawn up. The columns should be headed appropriately and any differing and/or similar features should be shown side by side.

Example: Tabulate the main differences between a man and a tree.

Man	Tree
Heterotrophic nutrition	Autotrophic nutrition
Locomotion	No locomotion
Growth is limited and occurs all over the body	Growth is unlimited and is restricted to certain regions
Responds quickly to brief stimuli	Responds slowly to long stimuli
No chlorophyll	Chlorophyll
No cellulose cell walls	Cellulose cell walls
No central vacuole in cells	Permanent central vacuole in cells
Compact body	Branched body

List. Write the relevant information briefly, numbering each fact 1 .... 2 .... 3 .... and so on. Questions which ask you to list the similarities and differences can be answered within a table, but do not list or tabulate the answer if the questions ask you to give an account.

Outline. Only the main facts are required and elaborate detail should not be given.

Compare. This should take the form of a written account which stresses the similarities and differences between the subjects given in the question.

Example: Compare the structure and function of an artery and a vein. For this question you should not write down all the characteristics of an artery followed by all the characteristics of a vein. Each point should be taken in turn and the common and differing features discussed, such as structure of the wall, the direction of blood flow, the blood pressure, the position in the body, and so on. Diagrams of sections through an artery and a vein should be drawn side by side to point out their similarities and differences.

Contrast. The approach is similar to that for comparisons but the emphasis should be on their differing features.

Compare and contrast. All the similarities between the subjects should be dealt with first, followed by their differences.

Distinguish between. This means that the differences between subjects must be shown. These questions should be dealt with in a similar way to 'contrast' type questions.

Discuss. Particular care should be taken with questions of this type. A full treatment is required but candidates often include much irrelevant information and are inclined to wander from the point. 'Discuss' frequently indicates a more critical approach and there is a possibility of expressing opinions. It is important to plan the answer, and as attempts at such questions often gain low marks they should not be attempted unless the topic is thoroughly known and understood.

Illustrated account. A diagram should be supplied, followed by a written account which refers to the parts labelled and particularly to their function.

Annotated diagrams. Large, clearly-labelled diagrams should be supplied with brief appropriate notes under each label. Unless specifically asked for in the question, additional writing underneath is not required.

Diagrams. Instructions at the beginning of examination papers may say that credit will be given for labelled diagrams where relevant. In some questions diagrams will be specified, so they should be practised as part of the revision. Where not specifically required they can be included to clarify the answer or to reduce the length of the written description. Information should not be duplicated. Diagrams should be large, with distinct, continuous outlines, and be clearly labelled. Labelling lines should be drawn with a pencil and ruler; they should touch the structures they are labelling, they should never cross, and they should be fairly evenly spaced around the diagram. Shading and colouring in should be avoided as they can obscure detail. Where colours or shading have been used, a key should be given unless the labels have explained their significance. Blood vessels need not be drawn with double lines if red (oxygenated blood) and blue (deoxygenated blood) are used with a key. All diagrams should be at least twice the size of those in this book, which have been reduced for space reasons. Drawing many identical structures should be avoided. For example, it is a waste of time to draw every organelle in a cell, only a few representative structures need be drawn very carefully.

Questions on experiments and practical work. The experiment is preferably written up under the headings – *Experiment, Method, Result, Conclusion*. Where appropriate a labelled diagram of the apparatus should be given. Your description should make clear any special features of the apparatus and experimental procedure, such as how often results are recorded, how long the experiment should be run for, and so on.

Preparation of questions. It is always advisable to briefly map out each answer before writing it in full. Frequently this procedure is neglected by candidates. With short-answer questions or where the question has been divided into parts it may not be necessary to plan the answer any further. However, in many cases, especially those questions starting with 'Discuss', 'Describe', 'Compare', or 'Contrast', the production of a brief outline can be very useful and removes the possibility of introducing irrelevances. It also helps the candidate to present the answer in a logical sequence and lessens the chance of leaving out important points.

Before the examination, planning questions for practice can be very useful, especially in view of the fact that many questions are repeated from previous years. If papers from previous years are obtained, specimen-answer plans can be mapped out.

Below is a very straightforward question and an outline answer.

Describe the composition and functions of blood

1. Composition

red cells } Structure (including shape);
white cells } size; number; formation;
platelets } lifespan
Plasma — Constituents

2. Function

a) Transport of:

oxygen   hormones
carbon dioxide   heat
soluble food   excretory products e.g. urea

b) Prevention of infection:

phagocytosis
antibodies (antitoxins)
blood clotting

Graphs: Many examining boards set questions on graphs and these can take two forms. Graphs may have to be constructed from data given, or questions may be asked on a graph which is provided.

Constructing graphs. The ability to do this is both necessary and advantageous, as many marks can be gained for technique and presentation, without specialist biological knowledge. Marks can be allocated for:

— full use of the graph paper
— selecting the correct axis for each set of figures
— dividing each axis equally into the correct number of units
— labelling each axis fully and accurately
— plotting each point with a small cross or a dot
— joining the points accurately
— giving the graph a correct title
— neatness and general presentation

Many candidates have difficulty in deciding the correct axis for each set of figures. Consider the following measurements of the growth of a human child:

Year	Height (cms)	Year	Height (cms)
0	54	11	142
1	76	13	152
5	108	15	169
9	132	19	175

The vertical axis is the y axis and the height (i.e. what the investigator finds out) is plotted on this axis. The horizontal axis is used for units which are determined by the investigator. In this case it is time in years.

Interpretation of graphs. Consider the graph below.

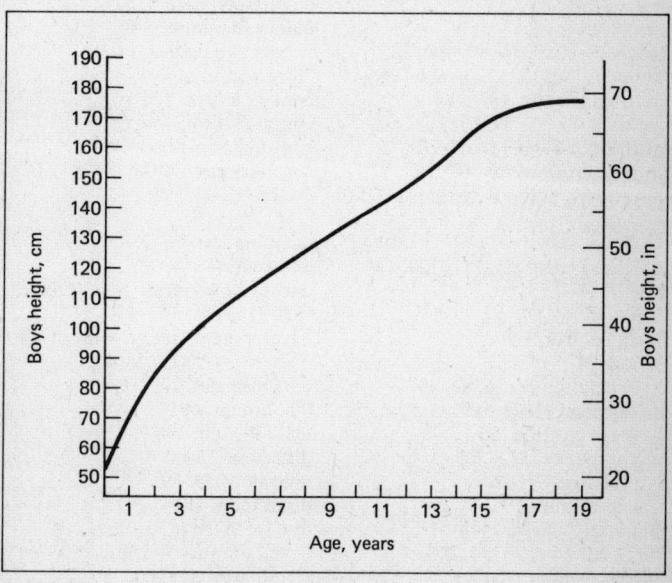

Figure 93. Growth of a typical boy

To answer questions such as 'What was the height of the child at 7', find 7 years on the x axis and with a ruler draw a line parallel to the y axis upwards from this point until it reaches the line of graph. From this point draw a line parallel to the x axis across to the y axis and where the line meets the y axis is the height required. Use a similar procedure to answer the question: 'How old was the child when he was 150 cms tall?'.

Index

abortion, 166
accidents, 192
accommodation (focusing), 124
active transport, 28, 95
adenosine triphosphate (ATP), 28, 29, 44
adrenalin, 69, 87, 114, 119
alcohol, 131, 132
allergy, 71, 103, 182
amino acids, 13, 40, 56
amniocentesis, 165
anoxia, 74
antibiotics, 190
antibodies, 47, 74, 80, 166, 182
antigens, 80, 182
antiseptics, 103, 182, 189
arteries, 72, 84, 89, 191, 217
arthritis, 145, 180
artificial respiration, 67, 193
autonomic nervous system, 114-115

bacteria, 104, 145, 150, 179, 190
balance (sense of), 111, 129, 144
bile, 51, 54, 55, 56, 116
birth, 165
birth control, 160
blood, 21, 72
   circulation of, 56, 82, 101
   clotting, 74, 77, 80
   detoxification, 56
   groups, 80, 172, 209
   pressure, 87, 88
   transfusion, 79
bone, 21, 42, 142, 144
bone marrow, 73, 74, 209
breast feeding, 166
breathing, 62, 119
bronchitis, 70, 179, 197

caffeine, 131, 132
calcium, 42, 44, 142, 144
cancer, 180, 191
capillaries, 53, 64, 72, 85
carbohydrates, 14, 40
carbon dioxide
   and gaseous exchange, 65
   from respiration, 29, 92, 96
   in photosynthesis, 205
   transport in blood, 76, 96, 145
carbon cycle, 205
cardiac muscle, 22, 86
cartilage, 21, 140
cell division, 31, 152
cell structure, 9
cellulose, 10
chlorophyll, 10, 204
chromosomes, 12, 151, 167, 176
circulatory system, 82
colour blindness, 175
connective tissues, 19, 98
constipation, 45
contact lenses, 126
contraception, 160
coordination, 106
coronary circulation, 84, 89
cretinism, 44, 118

deamination, 56
defaecation, 54, 92, 114
deoxyribose nucleic acid (DNA), 12, 15, 151, 167, 176, 213
development (see growth)
diabetes, 47, 117, 180
diagnosis, 88, 179
diaphragm, 60
diet, 39, 166, 196
diffusion, 27, 60
digestion, 34, 49
diphtheria, 184
disease, 179, 183
   and immunity, 189
   and medicine, 89, 189
   prevention of, 77, 199
   vitamin deficiency, 40
diuresis, 95, 118
drugs, 96, 130, 182, 187, 213

ears, 122, 126
endocrine system, 106, 114
energy in diet, 45
enzymes, 14, 24, 49, 52, 54, 97
epithelial tissues, 18

erythrocytes, 56, 73
evolution, 207
excretion, 56, 92
exercise, 69, 86, 145
external environment, 34, 181
eyes, 111, 121, 122

faeces, 54, 185, 211
fatigue, 145
fats, 15, 40, 53, 100
fertilization, 157
fibre (roughage), 45
fibrinogen, 73, 77
first aid, 193
fleas, 210
foetus, 163
food
   and diet, 39, 196
   chains, 204
   hygiene, 185, 189
   poisoning, 185, 187, 189
   preservation, 189
   tests, 48
footwear, 144
fractures (of bones), 145, 194

gametes, 151, 174
gaseous exchange, 64
genes, 167, 169, 208
genetics, 150, 167
German measles, 184
glucose, 14, 116
glycogen, 14, 55
glycolysis, 30
growth, 118, 135, 146, 150

haemoglobin, 13, 43, 65, 73, 76
haemophilia, 77, 176
hair, 99, 103, 104, 211
health and disease, 77, 179
health education, 199
hearing, 127
heart, 57, 72, 82, 119, 132
heart attacks, 89, 191
heart transplants, 89, 182
hepatic portal vein, 55
homeostasis, 57, 102
hormones, 51, 76, 106, 111, 114, 158, 164

houseflies, 185, 187
hygiene, 103, 184, 195
hypertension, 87

immunity, 79, 182, 189, 197
immunosuppression, 89, 182
implantation, 159
inflammation, 78
infection, 74, 77, 78, 90, 103, 166, 183, 189, 197
insomnia, 130, 132
insulin, 56, 116
iodine, 44, 118
iron lung, 67

kidney, 92, 182

labour (see birth)
lactation, 118
lactic acid, 30, 69
liver, 54, 184
long sight, 125
louse, 210
lungs, 35, 60, 70, 76, 92, 96
lymphatic system, 53, 75, 90
lymphocytes, 74, 90
lysosomes, 12

malaria, 185
meiosis, 152
menopause, 158
menstruation, 158
mental illness, 180, 192
metabolism, 23, 45, 97, 102, 132
micturition, 95, 114
milk, 47, 118, 166
minerals (in diet), 43, 54
mitochondria, 10
mitosis, 31, 150
muscles, 22, 106, 140
mutation, 151, 176, 208

National Health Service, 197
negative feedback, 58
nerve impulses, 107, 109, 119, 121
nervous system, 106
nervous tissue, 22, 107
nitrogen cycle, 206
nucleus, 10, 12

nutrition, 39

obesity, 47, 145
orgasm, 156
osmoregulation, 95
osmosis, 27
ossification, 144
ovaries, 118, 153
ovulation, 153
oxygen
    and gaseous exchange, 60, 65
    and respiration, 29
    transport in blood, 73, 76, 165

pacemaker (heart), 66, 86, 89
pancreas, 51, 116
pathogens, 74, 78, 180
penicillin, 190
phagocytosis, 29, 56, 74, 78
pituitary gland, 111, 118, 158
placenta, 81, 163, 165
platelets, 73, 74, 77
pollution, 70, 196
posture, 111, 142
pregnancy, 81, 159, 162
prenatal screening, 164
prophylaxis, 190
proteins, 13, 40, 166, 182
puberty, 146, 158

receptors, 66, 106, 113, 120
red blood cells, 56, 73
reflexes, 113
refuse disposal, 196
reproduction, 150
respiration (*see also* breathing)
    aerobic, 29, 33
    anaerobic, 30, 69
    artificial, 67, 193
Rhesus factor, 81, 171
ribosomes, 11, 167
rickets, 42, 180
roundworms, 212

safety (in the home), 192
saliva, 49, 54, 111
scurvy, 40, 180
sedatives, 131
sewage treatment, 181, 195, 200
sex chromosomes, 173

sex-linked inheritance, 175
short sight, 125
skeleton, 112, 135
skin, 77, 92, 96, 181, 189
sleep, 111, 130, 131
smell (sense of), 122
smoking and health, 70, 74, 191
spleen, 73, 74
spinal cord, 109, 112, 137
sterilization
    of food, 185, 189
    to prevent pregnancy, 161
stomach, 50
stimulants, 131
stroke, 89, 191
swallowing, 50, 61, 111, 212
sweating, 99, 101, 116
synapse, 109, 132

tapeworms, 211
taste (sense of), 122
teeth, 50, 104, 146
temperature control, 77, 96, 122
thyroid gland, 43, 118
tidal volume, 63
tissue fluid, 75, 78
tongue, 122
toxins, 78, 184, 190
tuberculosis (TB), 79, 181, 185
typhoid fever, 185
typhus, 181, 187, 210

umbilical cord, 90, 163
urea, 56, 76, 92, 94
urine, 76, 95, 156, 164

vaccination, 182, 189
vascular system, 35, 72
vectors (of disease), 184
veins, 72, 84, 217
venereal disease, 184, 188
ventilation, 60, 111
viruses, 181, 213
vision, 121, 122
vitamins, 40, 51, 54, 56, 122, 144

waste disposal, 196
water, 16, 43, 54, 92, 95, 102, 202
water pollution, 185
water purification, 202
white blood cells, 74, 78